Wind Tunnel and Propulsion Test Facilities

An Assessment of NASA's Capabilities to Serve National Needs

T0162223

Philip S. Antón, Eugene C. Gritton, Richard Mesic, Paul Steinberg

with Dana J. Johnson, Michael Block, Michael Brown,

Jeffrey Drezner, James Dryden, Tom Hamilton, Thor Hogan,

Deborah Peetz, Raj Raman, Joe Strong, William Trimble

Prepared for the National Aeronautics and Space Administration
and the Office of the Secretary of Defense

Approved for public release, distribution unlimited

NATIONAL DEFENSE RESEARCH INSTITUTE

The research described in this report was jointly sponsored by the National Aeronautics and Space Administration and the Office of the Secretary of Defense (OSD). The research was conducted in the RAND National Defense Research Institute, a federally funded research and development center supported by the OSD, the Joint Staff, the unified commands, and the defense agencies under Contract DASW01-01-C-0004.

Library of Congress Cataloging-in-Publication Data

Wind tunnel and propulsion test facilities : an assessment of NASA's capabilities to serve national needs / Philip S. Antón ... [et al.].
    p. cm.
    Includes bibliographical references.
    "MG-178."
    ISBN 0-8330-3590-8 (pbk. : alk. paper)
    1. United States. National Aeronautics and Space Administration—Evaluation. 2. Wind tunnels. 3. Aeronautics—Research—United States. 4. Airplanes—United States—Testing. 5. Airplanes—United States—Jet propulsion. I. Antón, Philip S.

TL567.W5W56 2004
629.134'52—dc22

                                                            2004014394

The RAND Corporation is a nonprofit research organization providing objective analysis and effective solutions that address the challenges facing the public and private sectors around the world. RAND's publications do not necessarily reflect the opinions of its research clients and sponsors.

RAND® is a registered trademark.

Cover design by Barbara Angell Caslon

Published 2004 by the RAND Corporation
1776 Main Street, P.O. Box 2138, Santa Monica, CA 90407-2138
1200 South Hayes Street, Arlington, VA 22202-5050
201 North Craig Street, Suite 202, Pittsburgh, PA 15213-1516
RAND URL: http://www.rand.org/
To order RAND documents or to obtain additional information, contact
Distribution Services: Telephone: (310) 451-7002;
Fax: (310) 451-6915; Email: order@rand.org

Preface

This monograph summarizes a one-year study (from June 2002 through July 2003) of the nation's wind tunnel and propulsion testing needs and the continuing ability that the National Aeronautics and Space Administration's (NASA's) major wind tunnel (WT) and propulsion test (PT) facilities[1] have in serving those needs; the study also identified any new investments needed and excess capacities within NASA. The study focused on the needs for larger (and, thus, more expensive to build and operate) test facilities and identified management issues facing NASA's WT/PT facilities.

This monograph should be of interest to NASA, the Department of Defense, the aerospace industry, the Office of Management and Budget, the Office of Science and Technology Policy, and congressional decisionmakers. Detailed supporting information for this monograph is contained in a longer companion technical report:

---

[1] Throughout this monograph, we use the term "WT/PT facilities" to mean wind tunnel facilities and propulsion test facilities, that is, the type of NASA facilities we assessed. Since individual facilities within this designation can be either wind tunnel facilities, propulsion test facilities, or both, "WT/PT facilities" serves as a generic term to encompass them all. That being said, when a specific facility is talked about, for clarity, we refer to it as a proper name and, if necessary, include its function (e.g., Ames 12-Foot Pressure Wind Tunnel). As well, the term "test facilities" and "facilities" can be substituted to mean "WT/PT facilities." Of course, NASA owns and operates other types of test facilities outside of WT/PT facilities, but our conclusions and recommendations do not apply to them.

Antón, Philip S., Dana J. Johnson, Michael Block, Michael Brown, Jeffrey Drezner, James Dryden, Eugene C. Gritton, Tom Hamilton, Thor Hogan, Richard Mesic, Deborah Peetz, Raj Raman, Paul Steinberg, Joe Strong, and William Trimble, *Wind Tunnels and Propulsion Test Facilities: Supporting Analyses to an Assessment of NASA's Capabilities to Serve National Needs*, Santa Monica, Calif.: RAND Corporation, TR-134-NASA/OSD, 2004 (referred to as Antón et al., 2004[TR], throughout the monograph).

The study was funded by NASA and jointly sponsored by NASA and the office of the Director, Defense Research and Engineering (DDR&E). It was conducted within the RAND National Defense Research Institute's (NDRI's) Acquisition and Technology Policy Center. NDRI is a federally funded research and development center sponsored by the Office of the Secretary of Defense, the Joint Staff, the unified commands, and the defense agencies.

Contents

Figures

Tables

Summary

This monograph reveals and discusses the National Aeronautics and Space Administration's (NASA's) wind tunnel and propulsion test facility management issues that are creating real risks to the United States' competitive aeronautics advantage.

Introduction

Wind tunnel (WT) and propulsion test (PT) facilities[1] continue to play important roles in the research and development (R&D) of new or modified aeronautic systems and in the test and evaluation (T&E) of developmental systems. The nation has invested about a billion dollars (an unadjusted total) in these large, complex facilities (some dating from the World War II era), which has created a testing infrastructure that has helped secure the country's national security and prosperity through advances in commercial and military

---

[1] Throughout this monograph, we use the term "WT/PT facilities" to mean wind tunnel facilities and propulsion test facilities, that is, the type of NASA facilities we assessed. Since individual facilities within this designation can be either wind tunnel facilities, propulsion test facilities, or both, "WT/PT facilities" serves as a generic term to encompass them all. That being said, when a specific facility is talked about, for clarity, we refer to it as a proper name and, if necessary, include its function (e.g., Ames 12-Foot Pressure Wind Tunnel). As well, the term "test facilities" and "facilities" can be substituted to mean "WT/PT facilities." Of course, NASA owns and operates other types of test facilities outside of WT/PT facilities, but our conclusions and recommendations do not apply to them.

aeronautics and space systems. Replacing many of these facilities would cost billions in today's dollars.

Many of these test facilities were built when the United States was researching and producing aircraft at a higher rate than it does today and before advances in modeling and simulation occurred. This situation raises the question of whether NASA needs all the WT/PT facilities it has and whether the ones NASA does have serve future needs. In fact, over the past two decades, NASA has reduced its number of WT/PT facilities by one-third. More recently, the agency has identified additional facilities that are now in the process of being closed. In addition, some of the remaining facilities are experiencing patterns of declining use that suggest they too may face closure.

As a result, Congress asked NASA for a plan to revitalize and potentially consolidate its aeronautical T&E facilities to make U.S. facilities more technically competitive with state-of-the-art requirements. Faced with Congress's request and with ongoing budgetary pressures from the Office of Management and Budget (OMB), NASA asked the RAND Corporation to clarify the nation's WT/PT facility needs and the continuing place that NASA's test facilities have in serving those needs, as well as to identify any new investments needed and excess capacities. NASA requested that RAND focus the study on the needs for large (and, thus, more expensive to build and operate) test facilities in six types of WT/PT facilities as well as to identify any management issues they face.

RAND conducted its research from June 2002 through July 2003. The study methodology involved a systematic review and analysis of national research, development, test, and evaluation (RDT&E) needs, utilization trends (historical and projected), test facility capabilities, and management issues. This analysis and its findings focused on answering four basic research questions:

1. What are the nation's current and future needs for aeronautic prediction capabilities, and what role do WT/PT facilities play in serving those needs?

2. How well aligned are NASA's portfolio of WT/PT facilities with national needs?
3. What is the condition (or "health") of NASA's portfolio of WT/PT facilities?
4. How should NASA manage its portfolio of WT/PT facilities?

Study Activities

To answer these questions, we conducted intensive and extensive interviews with personnel from NASA headquarters; personnel from NASA research centers at Ames (Moffett Field, Calif.), Glenn (Cleveland, Ohio), and Langley (Hampton, Va.), which own and manage NASA's WT/PT facilities; the staff of the Department of Defense's (DoD's) WT/PT facilities at the U.S. Air Force's Arnold Engineering and Development Center (AEDC, at Arnold AFB, Tenn.); selected domestic and foreign test facility owners and operators; U.S. government and service project officers with aeronautic programs; and officials in a number of leading aerospace companies with commercial, military, and space access interests and products.

In addition to RAND research staff, the study employed a number of distinguished senior advisers and consultants to help analyze the data received and to augment the information based on their own expertise with aeronautic testing needs and various national and international facilities.

Finally, the study reviewed and benefited from numerous related studies conducted over the past several years.

Perspectives on the Approach

The analytic method used in the study to define needs does not rely on an explicit national strategy document for aeronautics in general or for WT/PT facilities in particular because it does not exist. Lacking such an explicit needs document, we examined what categories of aeronautic vehicles the United States is currently pursuing, plans to pursue, and will likely pursue based on strategic objectives and current vehicles in use.[2]

Also, as *enabling infrastructures*, WT/PT facility operations are not funded directly by specific line items in the NASA budget.[3] The study's determination of WT/PT facility needs and the resulting conclusions and recommendations are therefore not based on the federal budget process as a direct indicator of policy dictates of facility need. We determined WT/PT facility need by identifying what testing capabilities and facilities are required given current engineering needs, alternative approaches, and engineering cost/benefit trade-offs. This, of course, can lead to a bias in the findings in that these assessments may be overly reflective of what the engineering field determines is important rather than what specific program managers are willing to spend on testing because of program budget constraints. Thus, when a needed facility is closed because of a lack of funding, there is a disconnect between current funding and prudent engineering need, indicating that the commercial and federal budget processes may be out of step with the full cost associated with research and design of a particular vehicle class and indicating a lack of addressing long-term costs and benefits.

Finally, while the study's focus was on national needs and NASA's WT/PT facility infrastructure, national needs are not dictated or met solely by NASA's test infrastructure; DoD, U.S. industry, and foreign facilities also serve many national needs. However, the study was *not* chartered or resourced to examine the sets of data for these alternative facilities to fully understand consolidation opportunities *between NASA and non-NASA WT infrastructures*. Based on our findings, such a broader study is important and warranted.

---

[2] Specific projects and plans were obtained from NASA, Office of Aerospace Technology (2001; 2002); NASA (2001; 2003); The National Aeronautics and Space Act of 1958; DoD (2000; 2002); FAA (2002); NRC (2001); Walker et al. (2002); NASA, Office of the Chief Financial Officer (n.d.); AFOSR (2002); and various DoD and commercial research and production plans.

[3] The *construction* of government WT/PT facilities are, however, very large expenditures that require explicit congressional funding, and certain facilities such as the National and Unitary facilities have associated congressional directives regarding operation and intent.

What Are the Nation's Current and Future Needs for Aeronautic Prediction Capabilities, and What Role Do WT/PT Facilities Play in Serving Those Needs?

Although applied aeronautics encompasses relatively mature science and engineering disciplines, there is still significant art and empirical testing involved in predicting and assessing the implications of the interactions between aeronautic vehicles and the environments through which they fly. Designers are often surprised by what they find in testing their concepts despite decades of design experience and dramatic advances in computer modeling and simulation known as computational fluid dynamics (CFD). This is, of course, especially true for complex new concepts that are not extensions of established systems with which engineers have extensive practical design and flight experience. But even improving the performance at the margin of well-established and refined designs—for example, commercial jet liners in areas such as reduced drag, fuel efficiency, emissions, noise, and safety (e.g., in adverse weather)—depends on appropriate and sufficient WT/PT facility testing.

Insufficient testing or testing in inappropriate facilities can lead to erroneous estimations of performance. Missed performance guarantees can impose extremely costly penalties or redesign efforts on airframe manufacturers, overly conservative designs from low estimations prevent trade-offs such as range for payload, and even a seemingly small 1 percent reduction in drag equates to several million dollars in savings per year for a typical aircraft fleet operator.[4]

For engineers to predict with sufficient accuracy the performance of their vehicles during design and retrofit, they need a range of capabilities, including high Reynolds number (Rn),[5] flow quality,

---

[4] See Mack and McMasters (1992) and Crook (2002).

[5] The Reynolds number is a nondimensional parameter describing the ratio of momentum forces to viscous forces in a fluid. The Mach number is a more familiar nondimensional parameter, describing the ratio of velocity to the sound speed in the fluid. When the flows around similarly shaped objects share the same nondimensional Rn and Mach parameters, the topology of the flow for each will be identical (e.g., laminar and turbulent flow distribution, location of separation points, wake structure), and the same aerodynamic co-

size, speed, and propulsion simulation and integration. As discussed below, these capabilities cannot be met by a single test facility but rather require a suite of facilities.

Also, *while CFD has made inroads in reducing some empirical test requirements capabilities, this technology will not replace the need for test facilities for the foreseeable future.* Flight testing complements but does not replace WT/PT facilities because of its high costs and instrumentation limitations.[6] The aeronautic engineering community does not have well-accepted handbooks of facility testing "best practices" or even rules of thumb from which to deduce testing requirements, nor is it possible from historical data to accurately predict returns on specific facility testing in terms of programmatic cost savings or risk reduction.

Thus, *aeronautic maturity does not nullify the need for test facilities but in fact relies on the availability and effective use of test facilities to provide important capabilities.* The nation continues to need general-purpose WT/PT facilities across all speed regimes, as well as for specialty tests. These facilities advance aerospace research, facilitate vehicle design and development, and reduce design and performance risks in aeronautic vehicles.

How Well Aligned Are NASA's Portfolio of WT/PT Facilities with National Needs?

NASA's WT/PT facilities are generally consistent with the testing needs of NASA's research programs, as well as with those of the broader national research and development programs. Currently, redundancy is minimal across NASA. Facility closures in the past decade have eliminated almost a third of the agency's test facilities in the categories under review in this study. In nearly all test categories,

---

efficients will apply (Batchelor, 1967). Airflow behavior changes nonlinearly and unpredictably with changes in Rn. Thus, it is important to test the flow conditions at flight (or near-flight) Rn to ensure that the flows behave as expected and that conditions such as undesired turbulence, separations, and buffeting do not occur.

[6] See, for example, Wahls (2001).

NASA has a single facility that serves the general- or special-purpose testing needs, although some primary facilities also provide secondary capabilities in other test categories. We found two noncritical WTs: (1) the Langley 12-Foot Subsonic Atmospheric WT Lab, which is redundant to the Langley 14×22-Foot Subsonic Atmospheric WT, and (2) the Langley 16-Foot Transonic Atmospheric WT, which is generally redundant to the Ames 11-Foot Transonic High-Rn and Langley National Transonic Facility WTs run in low-Rn conditions.

There are gaps in NASA's ability to serve all national needs. In most of these cases, though, DoD or commercial facilities step in to serve the gaps.

Finally, some of NASA's facilities that serve national needs have been or are in the process of being mothballed. While mothballing an important facility is preferred to abandonment, mothballing involves the loss of workforce expertise required to safely and effectively operate the facility. Thus, mothballing is not an effective solution for dealing with long periods of low utilization, and it puts facilities at risk.

What Is the Condition (or "Health") of NASA's Portfolio of WT/PT Facilities?

In looking at the condition, or health, of NASA's WT/PT facilities, two of the three key dimensions are (1) how *technically competitive* the facilities are and (2) how *well utilized* they are. Judged by those measures, NASA's portfolio is generally in very good condition. More than three-fourths of NASA's WT/PT facilities are competitive and effective with state-of-the-art requirements. In addition, more than two-thirds are well utilized. Overall, about two-thirds are both technically competitive and well utilized, with this number varying across the categories of test facilities.

However, there is room for improvement, especially in the high-Rn subsonic category and in reducing the backlog of maintenance and repair (BMAR) across NASA's portfolio. There also has been discussion in the testing community concerning both large and small

investments to improve NASA's test infrastructure, but it was difficult for our expert consultants and the user community to seriously
consider large investment candidates given declining budgets, facility
closures, and the failure of past efforts to obtain funding for facilities
with improved capabilities. Selected challenges, though, such as
hypersonic testing, will require additional research to develop viable
facility concepts for future investment consideration.

Finally, using a third dimension of health status—*financial
health*—we find that the full-cost recovery (FCR) accounting practice
imposed by NASA on the centers has serious implications for the
financial health of those facilities that are underutilized (about one-
third of the facilities in general, with variation across the test facility
category types). Average-cost-based pricing, decentralized budgeting,
poor strategic coordination between buyers and providers of NASA
WT/PT facility services, and poor balancing of short- and long-term
priorities inside and outside NASA are creating unnecessary financial
problems that leave elements of the U.S. WT/PT facility capacity
underfunded. With declining usage and FCR accounting, these
facilities run the risk of financial collapse.

How Should NASA Manage Its Portfolio of WT/PT Facilities?

NASA's primary management challenges break down into two questions. First, how can NASA identify the *minimum set* of WT/PT
facilities important to retain and manage to serve national needs?
Facilities that are in the minimum set should be kept, but those that
are not in the set could be eliminated (and, thus, constitute excess
capacity from a national strategic point of view). Second, what financial concerns and resulting management steps are needed to manage
the facilities within the minimum set?

Identifying the Minimum Set

Based on our analysis, 29 of 31 existing NASA WT/PT facilities constitute the minimum set of those important to retain and manage to

serve national needs. Thus, the test complex *within* NASA is mostly "right sized" to the range of national aeronautic engineering needs.

It is important to bear in mind that, while not the case within NASA, a few of NASA's facilities are redundant when considering the technical capabilities of the larger set of facilities maintained by commercial entities and by the DoD's AEDC. Whether these redundancies amount to the "unnecessary duplication" of facilities prohibited by the National Aeronautics and Space Act of 1958 was beyond the scope of this study. Further analysis of technical, cost, and availability issues is required to determine whether WT/PT facility consolidation and right-sizing across NASA and AEDC to establish a national reliance test facility plan would provide a net savings to the government and result in a smaller minimum set of WT/PT facilities at NASA.

Congress has expressed interest in collaboration between NASA and the DoD.[7] NASA and the DoD (through the National Aeronautics Test Alliance—NATA) have made some progress in their partnership,[8] but NASA's recent unilateral decision to close two facilities at Ames without high-level DoD review shows that progress has been spotty. Some in industry have expressed an interest in exploring collaborative arrangements with NASA and hope that this study will reveal to others in industry the risks to NASA's facilities and the need for industry to coordinate its consolidations with those of NASA and the DoD. Our study provides insights into the problem but offers only glimpses into the wider possibilities and issues surrounding broader collaboration.

Financial Support

The key management challenge remaining for NASA is to identify shared financial support to keep its minimum set of facilities from financial collapse given the long-term need for these facilities.

---

[7] See, for example, the GAO report on NASA and DoD cooperation entitled *Aerospace Testing: Promise of Closer NASA/DoD Cooperation Remains Largely Unfulfilled*, 1998.

[8] For example, NATA has produced a number of joint NASA and DoD consolidation studies. See NATA (2001a; 2001b; 2002).

In the extreme case at Ames, the lack of resident aeronautics research programs, combined with the center management's strategic focus toward information technology and away from ground test facilities, has left Ames WTs without support beyond user testing fees. Thus, Ames WT/PT facilities are vulnerable to budgetary shortfalls when utilization falls. Two Ames facilities that are unique and needed in the United States have already been mothballed and slated for closure as a result.

If NASA management is not proactive in providing financial support for such facilities beyond what is likely to be available from FCR pricing, then the facilities are in danger of financial collapse. In the near term, this market-driven result may allow NASA to reallocate its resources to serve more pressing near-term needs at the expense of long-term needs for WT/PT facilities. Given the continuing need for the capabilities offered by these facilities for the RDT&E of aeronautic and space vehicles related to the general welfare and security of the United States, the right-sizing NASA has accomplished to date, the indeterminate costs to decommission or eliminate these facilities, the significant time and money that would be required to develop new replacement WT/PT facilities, and the relatively modest resources required to sustain these facilities, care should be taken to balance near-term benefits against long-term risks. Collaboration, reliance, and ownership transferal options for obtaining alternative capabilities in lieu of certain facilities are possible, but even if these options are exercised, many NASA facilities will remain unique and critical to serving national needs. Key to subsequent analysis of these options is the collection and availability of the full costs of operating these facilities as well as the full costs associated with relying on alternative facilities.

Policy Issues, Options, and Recommendations

Table S.1 lays out and summarizes the main policy issues identified in the study along with the decision space for those issues and our assessment of the viability of those options. Nearly all options are specifically recommended or not recommended. One non-recommended option related to investments could be pursued, but it is unclear how viable it is in today's financial climate.

Note that the issues and options tend to be interrelated. For example, the determination of which facilities are important to keep is related to the question of what to do with low-utilization facilities. The recommended options are also related. For example, developing a long-term vision and plan for aeronautic testing, reviewing the technical competitiveness of facilities, and sharing financial support for facilities with users are interrelated.

The most critical issue is for NASA headquarters leadership to **develop a specific and clearly understood aeronautics test technology vision and plan**, to continue to support the development of plans to very selectively consolidate and broadly **modernize existing test facilities**, and to proscribe **common management and accounting directions** for NASA's WT/PT facilities. This vision cannot be developed in isolation from other critical decisions facing the nation. It must be informed by the aeronautic needs, visions, and capabilities of both the commercial and military sectors supported by NASA's aeronautical RDT&E complexes.

Given the inherent inability to reliably and quantitatively predict all needs for RDT&E to support existing programs much beyond a few months out, and the trends indicating a continuing decline in needed capacity to support these needs for the foreseeable future, **long-term strategic considerations must dominate**. If this view is accepted, then NASA must find a way to sustain indefinitely and, in a few cases, enhance its important facilities (or seek to ensure reliable and cost-effective alternatives to its facilities) as identified in this study.

Beyond this overarching recommendation, we propose the following, which reflect the entries shown in Table S.1:

- **NASA should work with the DoD to analyze the viability of a national reliance test facility plan,** since this could affect the determination of the future minimum set of facilities NASA should continue to support.

Table S.1
Key Policy Issues, Options, and Recommendations

Strategy	Technical Need	Facilities Needed	Operating Costs	Investments
Issue: How should testing be addressed strategically?	Issue: How should aeronautic testing needs be met?	Issue: Which facilities should NASA keep in its minimum set?	Issue: What to do with low-utilization facilities?	Issue: How should the facilities be funded?	Issue: What future investments should be made?
1. Develop a NASA-wide aeronautics test technology vision and plan **2. Work with the DoD to analyze the viability of a national test facility plan**	**1. Replace facility testing with CFD** **2. Replace facility testing with flight testing** 3. Use facility testing exclusively **4. Use appropriate mix of facility, CFD, and flight testing**	1. Keep all facilities that align with national needs **2. Leave out aligned facilities that are weakly competitive, redundant, and poorly utilized** 3. Leave out all facilities technically similar to those in the DoD	1. Close when utilization is low 2. Mothball when utilization is low **3. Reassess long-term needs and keep those strategically important** 4. Keep all facilities regardless of utilization	1. Recover all costs from users **2. Share support between users and NASA institutional funding** 3. Pay all costs through institutional funding so testing is free to users	**1. Eliminate backlog of maintenance and repair (BMAR)** **2. Conduct hypersonics test facility research** 3. *Pursue high-productivity, high-Rn subsonic and transonic facilities* **4. Continued investments in CFD research**

NOTES: **Bold = Recommended**; *Italics = Unclear*; Roman = Not Recommended.

- NASA should **continue to pursue all three complementary approaches—facility, CFD, and flight testing**—to meeting national testing needs. At this point, none can cost-effectively meet testing needs on its own.
- NASA should **identify the minimum set of test facilities important to retain and manage to serve national needs**. The facilities that do not belong in the minimum set are those that, despite their alignment with national needs, are weakly competitive, redundant, and poorly utilized. Further analysis of technical, cost, and availability issues are required to determine whether facility consolidation and right-sizing across NASA and the DoD would provide a net savings to the government.
- Utilization data is only one (nonexclusive) factor in determining what facilities to keep in the minimum set. In particular, utilization helps to decide what to do with redundant facilities. Thus, **poorly utilized facilities should be reassessed for strategic, long-term needs** rather than being eliminated out of hand; only those that do not survive that assessment are candidates for closure or mothballing. *Mothballing incurs the loss of important workforce expertise and knowledge.*
- NASA leadership should **identify financial support concepts to keep its current minimum set of facilities healthy for the good of the country**. It appears reasonable to ask users to pay for the costs *associated with their tests* (i.e., to pay for the short-term benefits), but forcing them to pay *all* operating costs (including long-term priorities such as the costs for facility time they are not using) through FCR *direct test pricing* (as is done at Ames) will further discourage use and endanger strategic facilities by causing wide, unpredictable price fluctuations in a world where government and commercial testing budgets are under pressure and are set years in advance. Also, we do not recommend setting the prices for user tests to zero because it closes one channel of information to users about the costs they are imposing on the infrastructure, can encourage overuse, and therefore cause limits on the availability of funding.

It is important to retain perspective on the magnitude of NASA's WT/PT facility costs relative to the investment value of the aerospace vehicles they enable or support. **While the approximate $125–130 million WT/PT operating cost is a significant sum, it pales in significance to NASA's overall budget of about $15,000 million**[9] **and the $32,000–58,000 million the United States invests in aerospace RDT&E each year.**[10] NASA should continue to closely reassess WT needs and ensure that excesses are not present. However, the agency should keep in mind the connection between these costs and the benefits they accrue. Engineering practices indicate that both the short- and long-term benefits are worth the cost in terms of the vehicles they enable, the optimization gains, and the reductions in risk to performance guarantees—even if short-term budgets are not currently sized to support the long-term benefits.

In terms of **investments**, we recommend the following:

- The $128 million **BMAR** at NASA facilities should be eliminated.
- Serious research challenges in **hypersonic air-breathing propulsion** may require new facilities and test approaches for breakthroughs to occur. This will require research in test techniques to understand how to address these testing needs and ultimately to look at the viability of hypersonic propulsion concepts being explored.
- To remain technically competitive, it makes sense to further consider investments in high-productivity, high-Rn subsonic and transonic facilities; however, current fiscal constraints make it unclear whether such investments should be pursued today.

---

[9] NASA 2002 Initial Operating Plan budget figures.

[10] Federal aerospace procurements and R&D expenditures in the period of FY1993–FY2001 ranged from a high of $58 billion in FY1994 to a low of $32 billion in FY2001 (Source: RAND research by Donna Fossum, Dana Johnson, Lawrence Painter, and Emile Ettedgui published in the *Commission on the Future of the United States Aerospace Industry: Final Report* [Walker et al., 2002, pp. 5–10]).

- Because of significant progress and utility, **continued investment in CFD** is recommended.

Beyond the general recommendations discussed above and highlighted in Table S.1, **NASA should focus on specific WT/PT facilities that require attention.** Financial shared support is most critical right now for the facilities at Ames: the 12-Foot, the National Full-Scale Aerodynamics Complex (NFAC), and the 11-Foot tunnels. Until an alternative domestic high-Rn subsonic capability can be identified, the Ames 12-Foot Pressure Wind Tunnel should be retained in at least a quality mothball condition. The NFAC is strategically important, especially for the rotorcraft industry, and needs immediate financial support to prevent the facility from abandonment at the end of FY2004. The Ames 11-Foot High-Rn Transonic facility currently provides excess capacity, but NASA should work with the DoD to establish long-term access to and clarified pricing for the AEDC 16T before considering whether to remove the Ames 11-Foot from the minimum set of needed facilities. Other facilities with unhealthy ratings also require attention, including the Langley Spin Tunnel, the Glenn 8×6-Foot Transonic Propulsion Tunnel, and the Ames 9×7-Foot Supersonic Tunnel.

In addition, **NASA should continue to explore options to preserve the workforce.** While our principal study focus has been on the test facilities themselves, these complex facilities become useless without trained personnel to operate them. Stabilizing NASA's institutional support for test facilities will help ensure that today's dedicated and competent workforce will be able to pass their skills on to future generations.

In conclusion, NASA has played critical roles in advancing the aeronautic capabilities of the country and continues to have unique skills important across the military, commercial, and space sectors —in terms of both research and support of our ability to learn about and benefit from advanced aeronautic concepts. Major wind tunnel and propulsion test facilities continue to have a prominent position in supporting these objectives. Unless NASA, in collaboration with the DoD, addresses specific deficiencies, investment needs, budgetary difficulties, and collaborative possibilities, the United States faces a real risk of losing the competitive aeronautics advantage it has enjoyed for decades.

Acknowledgments

We are very grateful for the strong support provided by the project sponsors: Dr. Jeremiah Creedon (NASA Associate Administrator for Aerospace Technology during the study), Dr. J. Victor Lebacqz (NASA Deputy Associate Administrator for Aerospace Technology during the study), Blair Gloss (NASA), Jean Bianco (NASA), Paul Piscopo (OSD/DDR&E), and Ted Fecke (OSD/DDR&E).

This study could not have been accomplished without the extensive support and insights provided by numerous officials and staff at the NASA Research Centers at Langley, Glenn, and Ames; NASA Headquarters; AEDC; the U.S. aerospace industry; and the test community in the United Kingdom.

Our team of senior advisers—H. Lee Beach, Jr., Eugene Covert, Philip Coyle, Frank Fernandez, Roy V. Harris, Jr., and Frank Lynch —provided very useful insights and guidance. Frank Lynch contributed many additional technical assessments on testing needs and facility capabilities. Gary Chapman (UC Berkeley) provided insights on computational fluid dynamics. Claire Antón offered insights into vehicle testing needs and NASA capabilities.

At RAND, Jerry Sollinger provided valuable structural insights into our charts and figures during the course of the study. Theresa DiMaggio, Maria Martin, Karin Suede, and Leslie Thornton gave their administrative support throughout the project. Phillip Wirtz edited the manuscript. Last but not least, we acknowledge the very valuable suggestions, questions, and observations from our reviewers, Frank Camm and Jean Gebman.

Abbreviations

16TT	16-Foot Transonic Tunnel
AEDC	Arnold Engineering and Development Center
AFB	Air Force Base
AFRL	Air Force Research Laboratory
AFSOR	Air Force Office of Scientific Research
AIAA	American Institute of Aeronautics and Astronautics
ATD	advanced technology demonstration
BMAR	backlog of maintenance and repair
CFD	computational fluid dynamics
CRV	current replacement value
DDR&E	Director, Defense Research and Engineering
DoD	Department of Defense
EOH	engine-on hours
FAA	Federal Aviation Administration
FCR	full-cost recovery
HYPULSE	Hypersonic Pulse
IT	information technology
JPL	Jet Propulsion Laboratory
JSF	Joint Strike Fighter
MER	Mars Exploration Rover

MOD	[United Kingdom] Ministry of Defence
MRTFB	major range and test facilities base
NASA	National Aeronautics and Space Administration
NATA	National Aeronautics Test Alliance
NFAC	National Full-Scale Aerodynamics Complex
NTF	National Transonic Facility
OMB	Office of Management and Budget
PT	propulsion test
R&D	research and development
RDT&E	research, development, test, and evaluation
RLV	reusable launch vehicle
Rn	Reynolds number
SLI	Space Launch Initiative
T&E	test and evaluation
TBD	to be determined
UAV	unmanned aerial vehicle
UCAV	unmanned combat aerial vehicle
UOH	user occupancy hours
WT	wind tunnel

Abbreviations

16TT	16-Foot Transonic Tunnel
AEDC	Arnold Engineering and Development Center
AFB	Air Force Base
AFRL	Air Force Research Laboratory
AFSOR	Air Force Office of Scientific Research
AIAA	American Institute of Aeronautics and Astronautics
ATD	advanced technology demonstration
BMAR	backlog of maintenance and repair
CFD	computational fluid dynamics
CRV	current replacement value
DDR&E	Director, Defense Research and Engineering
DoD	Department of Defense
EOH	engine-on hours
FAA	Federal Aviation Administration
FCR	full-cost recovery
HYPULSE	Hypersonic Pulse
IT	information technology
JPL	Jet Propulsion Laboratory
JSF	Joint Strike Fighter
MER	Mars Exploration Rover

MOD	[United Kingdom] Ministry of Defence
MRTFB	major range and test facilities base
NASA	National Aeronautics and Space Administration
NATA	National Aeronautics Test Alliance
NFAC	National Full-Scale Aerodynamics Complex
NTF	National Transonic Facility
OMB	Office of Management and Budget
PT	propulsion test
R&D	research and development
RDT&E	research, development, test, and evaluation
RLV	reusable launch vehicle
Rn	Reynolds number
SLI	Space Launch Initiative
T&E	test and evaluation
TBD	to be determined
UAV	unmanned aerial vehicle
UCAV	unmanned combat aerial vehicle
UOH	user occupancy hours
WT	wind tunnel

CHAPTER ONE

Introduction

Background

Wind tunnel (WT) and propulsion test (PT) facilities[1] continue to play important roles in the research and development (R&D) of new or modified aeronautic systems and in the test and evaluation (T&E) of developmental systems. The United States has invested about a billion dollars (an unadjusted total) in NASA's large, complex facilities (some dating from the World War II era[2])—investments that have created a testing infrastructure that has helped secure the country's national security and prosperity through advances in commercial

---

[1] Throughout this monograph, we use the term "WT/PT facilities" to mean wind tunnel facilities and propulsion test facilities, that is, the type of NASA facilities we assessed. Since individual facilities within this designation can be either wind tunnel facilities, propulsion test facilities, or both, "WT/PT facilities" serves as a generic term to encompass them all. That being said, when a specific facility is talked about, for clarity, we refer to it as a proper name and, if necessary, include its function (e.g., Ames 12-Foot Pressure Wind Tunnel). As well, the term "test facilities" and "facilities" can be substituted to mean "WT/PT facilities." Of course, NASA owns and operates other types of test facilities outside of WT/PT facilities, but our conclusions and recommendations do not apply to them.

[2] The Book Value of 26 of the 31 NASA facilities that fall within the scope of this study amounted to about $0.9 billion dollars based on data identified in the NASA Real Property Database. The book value is the simple sum of *unadjusted* dollars invested in past years in facility construction or modernization. Because, in many cases, decades have passed since construction, the book value is significantly lower than the cost it would take to build these facilities today.

and military aeronautics and space systems. Replacing these facilities would cost billions.[3]

Historically, the National Aeronautics and Space Administration (NASA) has owned a broad range of WT/PT facilities to support the nation's R&D testing needs.[4] NASA's WT/PT facilities are operated by both NASA personnel and contract labor. These facilities have supported NASA's own research, development, test, and evaluation (RDT&E) programs; supported industry's RDT&E needs; and provided secondary support to military RDT&E (when military research support provided benefits to the broader NASA and industry sectors).[5] While most NASA WT/PT facilities are used for NASA research, a few other NASA facilities are currently (but not exclusively) dominated by RDT&E related to commercial and military vehicles.

Many of the test facilities were built when the nation was researching and producing aircraft at a higher rate than it does today and before advances in modeling and simulation occurred. This situation raises the question of whether NASA needs all the WT/PT facilities it has and whether the ones it does have serve future needs. In fact, over the past two decades, NASA has reduced its number of WT/PT facilities by one-third. More recently, NASA has identified additional facilities that are in the process of being closed. In addition, some of the remaining facilities are experiencing patterns of declining use that suggest they too may face closure.

---

[3] The current replacement value (CRV) of 26 of the 31 NASA facilities that fall within the scope of this study totaled about $2.5 billion dollars based on data identified in the NASA Real Property Database. The CRV is derived by looking at similar types of buildings (e.g., usage, size) within the *Engineering News Magazine*'s section on construction economics. The magazine uses a 20-city average to come up with rough estimates of how much a building would cost to replace. Most NASA finance and facilities people believe that this average underestimates the actual cost of replacing WT/PT facilities, since they are more complex buildings than the "similar" building types available through engineering economics. Unfortunately, NASA has not found a better metric to compare buildings across the various field centers.

[4] See, for example, Baals and Corliss (1981) and Chambers (2000; 2003).

[5] Chambers (2000; 2003).

As a result, Congress asked NASA for a plan to revitalize and potentially consolidate its aeronautical T&E facilities to make U.S. facilities more technically competitive with state-of-the-art requirements.[6]

Objectives and Approach

Faced with Congress's request and with ongoing budgetary pressures from the Office of Management and Budget (OMB), NASA asked RAND to clarify the nation's WT/PT facility needs and the continuing place that NASA's test facilities have in serving those needs, as well as to identify any new investments needed and excess capacities. NASA asked the RAND Corporation to focus the study on the needs for large (and, thus, more expensive to build and operate) WT/PT facilities in the six categories shown in Table 1.1. NASA also asked RAND to identify any management issues these facilities face.

RAND conducted this study from June 2002 through July 2003. The study methodology involved a systematic review and analysis of national RDT&E *needs*, utilization *trends* (historical and projected), test facility *capabilities*, and *management issues*. This analysis and its findings focused on answering four basic research questions:

1. What are the nation's current and future needs for aeronautic prediction capabilities, and what role do WT/PT facilities play in serving those needs?
2. How well aligned are NASA's portfolio of WT/PT facilities with national needs?
3. What is the condition (or "health") of NASA's portfolio of WT/PT facilities?
4. How should NASA manage its portfolio of WT/PT facilities?

---

[6] U.S. Senate (2001, p. 108).

Table 1.1
Test Facility Categories for This Study

Test Facility Category	Mach Number Range[a]	Minimum Test Section Size[b]
Subsonic WT	0–0.6	6 feet
Transonic WT	0.6–1.5	4 feet
Supersonic WT	1.5–5.0	2 feet
Hypersonic WT	>5.0	1 foot
Hypersonic propulsion integration	>5.0	1 foot
Direct-connect propulsion	N/A	N/A

[a]Mach number is the ratio between the test speed and the speed of sound at the test conditions. Thus, a Mach number of 2.0 is twice the speed of sound, while a Mach number of 0.5 is half the speed of sound at test conditions.
[b]Nominally, test section size is the diameter of the test section perpendicular to the airflow direction. In wind tunnels where the vertical and horizontal dimensions are of similar magnitude yet differ (e.g., 9 feet high and 15 feet wide), we considered the largest dimension against this criterion.

Study Activities

To answer these questions, we conducted intensive and extensive interviews with personnel from NASA headquarters; personnel from NASA research centers at Ames (Moffett Field, Calif.), Glenn (Cleveland, Ohio), and Langley (Hampton, Va.), which own and manage NASA's WT/PT facilities; the staff of the Department of Defense's (DoD's) WT/PT facilities at the U.S. Air Force's Arnold Engineering and Development Center (AEDC, at Arnold AFB, Tenn.); selected domestic and foreign test facility owners and operators; U.S. government and service project officers with aeronautic programs; and officials in a number of leading aerospace companies with commercial, military, and space access interests and products.

We used three semistructured interview protocols to provide advanced notice of the study needs and a level of consistency across the interviews. First, we used an interview protocol for our initial on-site visits and discussions with NASA programs, test facility managers, and DoD users. Second, we developed a questionnaire to solicit projected utilization of NASA facilities. Finally, we used detailed supplementary questionnaires to (1) solicit additional insights from aerospace vehicle designers in industry and the DoD about their strategic needs in each of the six WT/PT facility categories shown in Table 1.1 and (2) probe their preferred facilities and acceptable or possible alternatives, the bases being used for facility selections (technical, business environment, etc.), their needs for new facilities, and their assessments of computational fluid dynamics' (CFD's) role in reducing facility requirements. The questionnaires are provided in Appendix D of Antón et al. (2004[TR]).

In addition to RAND research staff, the study employed a number of distinguished senior advisers and consultants to help analyze the data received and to augment it based on their previous expertise with aeronautic testing needs and various national and international facilities.

Finally, the study reviewed and benefited from numerous related studies conducted over the past several years.

Perspectives on the Approach

The analytic method used in the study to define needs does not rely on an explicit national strategy document for aeronautics in general and for WT/PT facilities in particular because it does not exist. Lacking such an explicit needs document, this study examined what categories of aeronautic vehicles this country is currently pursuing, plans to pursue, and will likely pursue based on strategic objectives and current vehicles in use.[7] In some cases, no explicit vehicle planning exists, but the study assessed current uses and determined that future vehicles will need to be produced in the future. For example, we assumed that the United States would continue to need commercial and military rotorcraft and military bomber vehicles despite the lack of a strategic document on committing the resources of the country to their RDT&E.

---

[7] Specific projects and plans were obtained from NASA, Office of Aerospace Technology (2001; 2002); NASA (2001; 2003); The National Aeronautics and Space Act of 1958; DoD (2000; 2002); FAA (2002); NRC (2001); Walker et al. (2002); NASA, Office of the Chief Financial Officer (n.d.); AFOSR (2002); and various DoD and commercial research and production plans.

Despite the existence of planning documents that discuss future vehicles, none of the planning documents explicitly discusses WT/PT facilities. Thus, this study used the vehicle categories as the basis for an examination of test facility capabilities needed for RDT&E of those vehicles. This analysis examined engineering design practices as evidenced by expert analysis, advocacy, and survey responses from the research and design communities. Thus, national needs for WT/PT facilities are traced back to the vehicles they enable. If strategic decisions are made in the future that these vehicles are no longer needed, then the results of this study can be used to understand what facilities are no longer needed. For example, if the DoD and commercial sectors decide that rotorcraft are no longer important, then the WT/PT facility needs that support rotorcraft RDT&E can be eliminated. However, lacking an explicit strategic policy decision that the country will no longer pursue rotorcraft, this study included these needs in the analysis and conclusions. The study does not dictate what vehicles the country should produce; it merely maps what WT/PT facilities the country needs based on the vehicles in evidence that the country is pursuing and apparently will still need based on a review of existing planning documents and strategic positions.

Note also that as *enabling infrastructures*, WT/PT facility operations are not funded by specific line items in the NASA budget.[8] NASA facilities are not line items in the congressional budget requiring explicit congressional policy directives about these facilities' needs. The study's determination of these needs and the resulting conclusions and recommendations are therefore not based on the federal budget process as a direct indicator of policy dictates of WT/PT facility need. Since WT/PT facilities are enabling infrastructure for vehicle categories that enter such policy debates, the study focused on those vehicle categories and the pursuits of such vehicles as the bases of engineering analysis. Policies will dictate specific vehicle productions over time in the future; this study addresses what facility capa-

---

[8] The *construction* of government WT/PT facilities are, however, very large expenditures requiring explicit congressional funding, and certain facilities such as the National and Unitary facilities have associated congressional directives regarding operation and intent.

bilities will enable the United States to produce such vehicles when such policies arise.

Moreover, the study viewed NASA and Congress's request for an assessment of WT/PT facility needs as an opportunity to inform budget decisions rather than a dictate to explain the needs as evidenced by current policy budgetary decisions.

The analytic method used in this study defines the specific WT/PT facility needs identified in the areas of national security, research, development, production, and sustainment as those required to enable the prudent research, design, and testing of vehicles classes of interest to the United States. Test facility need was determined by engineering practices to research new aeronautic concepts, explore and select new designs, and validate performance. The aeronautic experts consulted in the approach applied their best judgment on what testing capabilities and facilities are required given current engineering needs, alternative approaches, and engineering cost and benefit tradeoffs. These descriptions of needs reflected current and anticipated approximations that are being explored and used to keep WT/PT facility testing to a minimum, but they do not necessarily reflect short-term budgetary pressures within programs. They are the best judgments of the engineering community about what is needed strategically to research and produce the next generation of aerospace vehicles in all classes.

This, of course, can lead to a bias in the findings because these assessments may overly reflect what the engineering field determines is important rather than what specific program managers are willing to spend on testing because of program budget constraints. For example, the study findings point to a disconnect between current funding and prudent engineering need. Future utilization levels may not reflect the engineering assessments if future disconnects remain. Also, the study found in certain places that underfunding of programs has driven those programs to use facilities that are not appropriate to meet their needs—facilities that represent shortfalls or insufficient compromises rather than prudent capability choices in a market.

The disconnect may also indicate that the commercial and federal budget processes may be out of step with the full cost associated with the research and design stages of a particular vehicle class. If, in the extreme case, this process reaches the point in which the federal government decides it can no longer afford to pursue entire vehicle classes both now and in the long term, then the results of this study can be used to indicate which WT/PT facilities are therefore no longer needed.

Study Scope

While the study focus was on national needs and NASA's WT/PT facility infrastructure, national needs are not dictated or met solely by NASA's test infrastructure; DoD, U.S. industry, and foreign facilities also serve many national needs. Therefore, the study analyzed potential consolidation opportunities *within NASA's WT/PT facility infrastructure* and technical considerations for key non-NASA facilities that might alternatively serve national needs. RAND collected data on and analyzed selected DoD and foreign WT/PT facilities to understand the breadth, depth, and quality of these facilities that are similar to NASA's and developed a base of knowledge for addressing the competitive need for revitalizing existing NASA facilities. However, the study was *not* chartered or resourced to examine the sets of data for these alternative facilities to fully understand consolidation opportunities *between NASA and non-NASA WT/PT facility infrastructures*. Based on our findings, such a broader study is important and warranted.

Also, the study reviewed high-level cost data available from NASA on its WT/PT facility operations but did not engage in an in-depth analysis of the different accounting standards and policies in use at the NASA centers. Moreover, NASA was implementing full-cost accounting while this study was being conducted; as a result, some cost data were in preparation and, therefore, not available.

Organization of This Monograph

Each of the following chapters addresses one of the four major research questions posed above and summarizes the results of our analyses to address those questions. More specifically, each chapter opens with a question, provides the key summary answer to the question, and then provides supporting information for the answer. Additional detailed information for the conclusions and recommendations can be found in the companion technical report (Antón et al., 2004[TR]).

Chapter Two focuses on national WT/PT facility needs; Chapter Three looks at which of NASA's existing test facilities serve national needs; Chapter Four examines the condition of NASA's WT/PT facilities; and Chapter Five explores management issues facing NASA's test facilities. Chapter Six summarizes the study's conclusions and recommendations.

What Are the Nation's Current and Future Needs for Aeronautic Prediction Capabilities, and What Role Do Wind Tunnel and Propulsion Test Facilities Play in Serving Those Needs?

One argument that has gained currency is that the nation really does not have much need for aeronautic prediction capabilities. This argument is driven by the fact that applied aeronautics encompasses relatively mature science and engineering disciplines. Thus, the argument goes, while such prediction capabilities were obviously invaluable in the early days of aeronautics, they are less useful today. A secondary argument, related to the first, is that for whatever aeronautic prediction needs the country does have, the United States does not need wind tunnel and propulsion test facilities to meet those needs, especially given the existence of flight testing and the rise and growing sophistication of simulation technology like computational fluid dynamics.

Based on our research, we find that, **despite overall declines in aerospace research and aerospace vehicle production rates, the nation continues to pursue performance improvements in past aerospace vehicles types while exploring new vehicles and concepts, resulting in demands for aeronautic prediction capabilities.** This demand cuts across all categories of need—strategic, research, development, production, and sustainment—for all speed regimes, and for specialty tests to advance aerospace research and to reduce the risk in developing aerospace vehicles. As for the continuing need for WT/PT facilities, we further find that, **although CFD has made inroads in reducing some empirical test simulation requirements, CFD will not replace the need for test facilities for the foreseeable future.**

Moreover, because of high costs and instrumentation limitations, flight testing complements but does not replace WT/PT facilities.

The remainder of this chapter provides support for these findings, starting with how such needs can be met.

How Can Aeronautic Prediction Needs Be Met?

Three approaches are available to predict aeronautic behavior: WT/PT facilities (i.e., empirical testing in controlled ground facilities), CFD, and flight testing. Let us first consider the two alternatives to WT/PT facilities.

CFD has made inroads in reducing *some* empirical test simulation needs, but the technology will not eliminate the need for test facilities for the foreseeable future.[1] Estimates of the time frame for computer simulation to be capable of fully replacing WT/PT facility testing are on the order of decades.[2] This diminishes neither the importance of CFD nor the need for continued investment in simulation technology, but it puts the capability in proper perspective as a *complementary* resource to ground test facilities and flight testing.[3] To that end, CFD has proven an excellent tool for preliminary design configuration screening (for example, simulation of conventional aircraft at cruise condition has allowed for up to 50 percent reductions

---

[1] See, for example, Rahaim et al. (2003) for a good overview of the status and future plans for CFD, as well as Oberkampf and Blottner (1998) for a broad survey-level discussion of the ways in which CFD can encounter inaccuracies.

[2] The *DoD Aeronautical Test Facilities Assessment* (1997) said that "extensive use of CFD to replace wind tunnel data [is] 20 to 40 years away," and other expert assessments contended that this may be an underestimate.

[3] See Giunta, Wojtkiewizc, and Eldred (2003) for modern design-of-experiment methods for CFD codes. See also Streett (2003) and Streett et al. (2003) for good examples of experiments that blend CFD and other experimental techniques.

in physical testing at the screening stage[4]). For an extremely limited set of cases, it is even possible to make predictions more accurately than with WT/PT facilities.[5] *However, validated CFD capability is limited to these relatively simple flow conditions. The technology is not yet considered reliable for predicting the characteristics of the complex separated flows that dominate many critical design points for an aircraft.* Continued investment in CFD should result in steady advancements in the envelope of validated simulation capabilities, but the validation process itself will require many precise needs for WT/PT facilities experiments.[6] Ironic as it may be, we cannot hope to eventually replace the need for WT/PT facilities without maintaining high-quality WT/PT facilities in the meantime.

Flight testing plays a dominant role during final refinement, validation, and safety verification of a production aircraft, but except for small vehicles for which multiple, full-scale, flight-capable vehicle concepts can be quickly and relatively inexpensively produced (or for some selected vehicles for which engineers cannot economically develop a ground test facility), flight testing complements but does not replace WT/PT facility testing.

Therefore, for most aeronautic testing, WT/PT facilities remain the most viable and cost-effective tools for predicting aeronautic behavior.

---

[4] Screening-stage reductions were cited by multiple industry design experts in response to our survey questions. See also Beach and Bolino (1994), Crook (2002), and Smith (2004) for additional discussions on the effects of CFD testing on WT/PT facility testing hours. However, the benefits of using CFD for initial screening and to improve testing efficiency do not necessarily indicate a reduction in *overall* wind tunnel testing hours. Rather, a complementary CFD program presents an opportunity to shift more testing resources from preliminary explorations to final optimization. Respondents made it very clear that decisions on quantity of testing are primarily budget-driven, and they will test as much as they can afford to address the range of technical concerns and reduce important risks when possible.

[5] See Oberkampf and Aeschliman (1992) and Walker and Oberkampf (1992).

[6] Validation challenges include knowing a great deal more about the flow field in the tunnel than at the surface of the model, significant instrumentation, tests with multiple model sizes, and significant funding. See, for example, Aeschliman and Oberkampf (1998).

What Aeronautic Prediction Capability Needs and Requirements Exist?

When it comes to determining WT/PT facility needs, it is critical to realize that such needs and requirements mean different things to different people (e.g., those in program offices, at NASA Headquarters, in OMB, on congressional staffs, in industry). For this study, we found it useful to define and assess three types of requirements and needs: (1) development, production, and sustainment requirements; (2) research requirements; and (3) strategic needs.

Development, Production, and Sustainment Requirements

Development, production, and sustainment requirements address planned and programmed vehicle and component programs and activities that are "on the books." They can be identified in specific, technical capabilities that directly relate to services provided by NASA's test infrastructure.

Vehicle Types and Production Rates. The argument that we do not require that much aeronautic prediction capacity is driven mostly by the sense that development and production activities are declining. And, in fact, Figure 2.1 (which plots our count of the number of new aircraft designs reaching first flight per decade) shows that the number of new aerospace vehicles put into production has indeed decreased today from historic highs. These numbers reinforce what has been generally expressed in the aeronautic community—that fewer vehicles are being put into production today than in the past.

As shown in the figure, the number of civilian aircraft starts has shrunk from about eight per decade in the 1950s to about one per decade in the 1990s and in the current decade. Military aircraft starts have also slowed (especially when compared with the 1950s). However, the nature of current vehicle starts is also changing. Manned military aircraft programs are larger and more complex than their predecessors, while unmanned aircraft are becoming the largest part of the military aircraft starts.

Figure 2.1
Number of New Aircraft Designs Reaching First Flight: 1950–2009 (estimated)

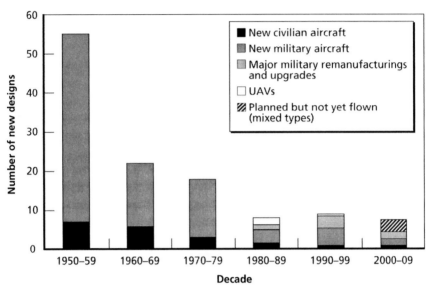

SOURCES: AAI (2000); Airborne Laser (2002); Boeing (1997, 2001, 2002a, 2002b, 2003); Corliss (2003); Drezner and Leonard (2002); Drezner et al. (1992); General Atomics (2002); GlobalSecurity.org (2002a, 2002b); Lockheed Martin (2001, 2003); Lorell and Levaux (1998); Northrop Grumman (2003); Pioneer UAV Web site; Raytheon (1998).
RAND MG178-2.1

While the figure does show a rapid decline, it also helps to make a more subtle point: *No vehicle classes have been eliminated from future needs, and each class will continue to require empirical prediction of air-flow behavior across a range of design considerations.* Even beyond the existing programs, it is clear that the country will eventually need to produce each existing vehicle class. For example, a new bomber will sooner or later be produced even though no current programs are planned, and the Army and commercial industry will not forgo rotor-craft. Thus, *the aeronautic prediction capabilities required to produce these vehicles (no matter their production rate) must be preserved, or the country will risk our ability to produce them.* When redundancy is eliminated, utilization reflects *management challenges* to keep low-use facilities healthy for future needs, given low revenues from testing rather than a metric of the *number* of facilities of each type that the country needs.

Furthermore, the mix of vehicles being explored is expanding; therefore, the United States will need to satisfy the state-of-the-art aeronautic prediction capability needs emerging from these new vehicle types.

Test Facility Roles in Predicting Vehicle Performance. Although applied aeronautics encompasses relatively mature science and engineering disciplines, there is still significant art and empirical testing involved in predicting and assessing the implications of the interactions between aeronautic vehicles and the environments through which they fly. This is, of course, especially true for complex new concepts that are not extensions of established systems that engineers have extensive practical design and flight experience with. But even improving the performance at the margin of well-established and refined designs—for example, commercial jetliners in areas such as reduced drag, fuel efficiency, emissions, noise, and safety (e.g., in adverse weather)—depends on appropriate and sufficient WT/PT facility testing.

Insufficient testing (or testing in inappropriate facilities such as those with insufficient Rn or flow quality) can lead to erroneous estimations of performance. Missed performance guarantees can impose extremely costly penalties or redesign efforts on airframe manufacturers, overly conservative designs from low estimations prevent trade-offs such as range for payload, and even a seemingly small 1 percent reduction in drag equates to several million dollars in savings per year for a typical aircraft fleet operator.[7]

The fact that aeronautic R&D involves a fair amount of "cut and try" artistry makes the general and continuing need for aerodynamic testing clear, but it does not tell us about how valuable specific test facilities might be. For engineers to predict with sufficient

---

[7] See Mack and McMasters (1992) and Crook (2002).

accuracy the performance of their vehicles during design and retrofit, they need a range of capabilities needed, including high Rn, flow quality, size, speed, and propulsion simulation and integration. As discussed below, these capabilities cannot be met by a single test facility but rather require a suite of facilities. Thus, *aeronautic maturity does not nullify the need for test facilities but in fact relies on the availability and effective use of test facilities to provide important capabilities.*

Research Requirements

Research requirements address those R&D activities associated with advancing the aeronautics state of the art (and, thus, activities associated with supporting the U.S. aerospace industrial base and academia).

Research requirements address developing new systems concepts and technology, modernizing and extending the capabilities of the test infrastructure itself, and making advances in aerospace sciences and engineering (e.g., flow physics theory, materials and structures, propulsion, systems integration, and CFD).

Table 2.1 lists many specific advanced vehicle concepts being considered and the time frame in which RDT&E for them may start. The table shows that there are numerous concepts in the pipeline that will require aeronautic prediction capability in the future.

Strategic Needs

Strategic needs derive from long-term national goals and strategies, such as "national security" (writ large). The need to ensure that national security–related programs do not depend on sources beyond our national control (e.g., foreign companies or governments) for critical functions (e.g., developing and testing the next-generation fighter aircraft) is one possible example of a strategic need we considered. We also assumed that national security involves the need for a robust, competitive national industrial base, including healthy commercial aeronautical and space capabilities.

Table 2.1
Aerospace Vehicle Concepts Through 2020

Program	Description	Status	Half-Decade Time Frame
Military Aircraft			
Unmanned combat aerial vehicle (UCAV) (X-45)	First generation, subsonic	Advanced technology demonstration (ATD) ongoing	<2005
UCAV-N [Navy] (X-47)	First generation, subsonic	ATD ongoing	<2005
UCAV	Supersonic	Concept	2006–2010
UCAV	Hypersonic	Concept	2011–2015
Tanker	B-767 conversion	Planned	<2005
Quad tilt-rotor	Short takeoff and vertical landing (STOVL) requirement	Concept	2006–2010
Tactical aircraft	C-130 replacement	Analysis of alternatives	<2005
F/A-18G Growler	Electronic warfare aircraft, Navy	In planning	<2005
Multi-Mission Maritime Aircraft (MMA)	P-3 replacement	In planning	2006–2010
Multi-Sensor Command and Control Aircraft (MC2A)	Air Force bomber	In planning	2006–2010
FB-22	F-22 modification for long-range strike	Concept	2006–2010
Long-range strike	Air Force bomber	Concept	2011–2015
Gunship	AC-130J follow-on	Concept	2006–2010
YAL-1A	Airborne laser	CTD	<2005
Quiet supersonic platform		Tech	2006–2010
Sensor craft	Sensor embedded in structure	Concept	2011–2015
HEL fighter	High-energy laser	Concept	2006–2010
HPM UCAV	High-power microwave	Concept	2006–2010
M-X	Special Operations Forces insertion mission	Concept	2006–2010
KC-X	Tanker/cargo high-efficiency design	Concept	2011–2015
Advanced Tactical Transport	Intra-theater transport	Concept	2006–2010

Table 2.1—Continued

Program	Description	Status	Half-Decade Time Frame
Military Aircraft (cont.)			
X-50A C/RW	Conard/rotor wing tech demo	ATD ongoing	<2005
Pelican	Very large transport; flies in ground effect		2006–2010
Missiles			
Cruise missile	Subsonic, extended range	Concept	<2005
Cruise missile	Supersonic	Concept	2006–2010
Cruise missile	Hypersonic	Concept	2011–2015
HyShot	Scramjet engine	ATD	<2005
SHOC (standoff high-speed option for counter-proliferation)	Mach 3.5–4.5 missile with 400–600 nmi range	ACTD FY04	<2005
ERAM	Extended range active missile for Navy	Concept	<2005
Space			
FAAST	Access to space, future strike		2011–2015
Trans-atmospheric vehicle		Concept	2016–2020
Hypersonic vehicle		Concept	2016–2020
2nd-generation reusable launch vehicle (RLV)	Space Launch Initiative (SLI) program	Tech	<2005
3rd-generation RLV		Concept	2011–2015
X-37	NASA, Boeing (SLI)	Tech demo	<2005
X-43A			2006–2010
X-43B			2011–2015
X-43C			2011–2015
Military space plane	Single stage to orbit or two stage to orbit	Concept	2016–2020
Commercial Aircraft			
Exec-u-jet	Supersonic business jet	Tech	2006–2010
7E7	Boeing commercial jet	Development	<2005

Strategic needs also include those that are driven by conceptual systems and activities flowing from long-term plans, visions, and research initiatives. We postulated that they should also include maintaining test facility strategic reserves to hedge against such uncertainties as vehicle or test facility accidents, system design flaws, unplanned system modifications, and policy changes. Given the relatively long timelines associated with developing or reconstituting test facilities to meet unexpected needs (e.g., it nominally takes 10 years to construct a new facility[8]), we postulated that there is clearly a need to maintain a prudent reserve.

Ultimately, long-term strategic visions are realized in actual vehicle concepts (which generate research needs) and programs (which generate development, production, and sustainment needs). The research, design, certification, production, sustainment, and redesign of these vehicles produce more specific near-term needs for test facilities. Anticipated strategic needs include the following:

- **space access**—subsonic through hypersonic speeds and recent efforts for air-breathing hypersonic propulsion
- **commercial transports**—subsonic through supersonic and their propulsion systems
- **military vehicles**—subsonic through supersonic vehicles and their propulsion systems
- **military weapons**—supersonic through hypersonic missiles and their propulsion systems.

As discussed below, within each of these three needs categories, needs can be specified in generic terms by flight regime (i.e., subsonic, transonic, supersonic, or hypersonic) and by unique, special-purpose applications (e.g., icing, acoustics, emissions) through analysis of the vehicle's speed range as well as specific operating challenges and goals (e.g., whether the vehicle must operate under icing condi-

---

[8] Based on data obtained from AEDC (see Antón et al., 2004[TR]).

tions, whether acoustic levels are an issue for operation at specific airports).

Needs also differ by user sector. DoD and commercial users have distinct needs for NASA support that are determined by their particular missions and objectives. Specifically, NASA focuses on space RDT&E and basic aeronautics R&D;[9] the DoD concentrates on the aeronautic vehicles in support of the warfighter; and industry is driven by market forces in the commercial airline and business aircraft sectors and by shareholder interests. Broadly speaking, NASA tends to focus on an R&D perspective, while the DoD and industry are driven by development, testing, and evaluation of programs. The result is a different approach to testing objectives, processes, and procedures that, in turn, influences the kinds of programs using specific WT/PT facilities.

In all cases, specifying facility testing needs as much as possible in quantitative terms[10] helps in addressing capacity questions (i.e., How many facilities of a certain type does the country need?). Unfortunately, programmatic needs are often uncertain beyond the near term (i.e., beyond a few years at best, and weeks to months at worst). Also, qualitative characteristics of WT/PT facilities (e.g., responsiveness; flexibility; availability of on-site aerospace expertise; productivity focus; and user confidence in, and familiarity with, test facilities, databases, and operations) accentuate the unique technical

---

[9] NASA's current aeronautics testing needs are focused in the next three to five years on making pragmatic improvements in fuel efficiency, acoustics, and engine performance, as well as on exploring new concepts for access to space and hypersonic missiles. Longer-term concepts, as described in NASA's Aeronautics Blueprint, include ideas for new aerodynamic aircraft concepts (blended wing, morphing wing, active flow control, etc.), but these concepts are being pursued at preliminary levels of support and funding, with little attention being paid to practical issues that currently have no known solutions but must be addressed to transition those concepts to production. The Blueprint even presents some novel future concepts, such as personal air vehicles that could be explored in 20 to 30 years as technological advances in semiautomated control are made in small, unmanned air vehicles, air traffic control, and other efforts. These concepts may drive future testing needs if breakthroughs and funding are realized and practical production and operational issues can be addressed. See NASA Office of Aerospace Technology (2002).

[10] For example, in estimates of user occupancy hours (UOH) for each class of wind tunnel needed, or engine-on hours (EOH) for each type of propulsion test facility needed.

characteristics of each facility and make it more difficult to quantitatively measure facility needs.

How Do Strategic Needs Match Up Against WT/PT Facility Types?

Given that WT/PT facilities will remain critical tools for meeting aeronautic prediction needs, we sought to determine how the six types specified in Table 1.1 match up against strategic needs informed by R&D, production, and sustainment need categories that are long term and continuing. We determined strategic test facility needs from reviews of engineering requirements for each type of facility, using RAND staff assessments, expert consultants from the aeronautic design and test communities, discussions with government users, and a semiformal survey of NASA researchers and the leading aerospace designers in industry and the DoD, reaching 24 different aerospace organizations.

Given strategic vehicle needs, technical aeronautic considerations, and discussions and surveys of facility users, what general kinds of WT/PT facility types does the nation need? Since these types of facilities cost from hundreds of millions to billions of dollars and average 10 years to construct, we took the longest view possible in summarizing national needs, being informed by but also looking beyond cyclic utilizations, program starts (and ends), and even midterm strategic plans that only last a few years.

Test facilities can be grouped into two broad categories along the lines of general- and special-purpose test needs, which we briefly discuss below in terms of the six categories of facilities identified above.

General-Purpose Subsonic, Transonic, and Supersonic WTs
General-purpose subsonic, transonic, and supersonic WTs have broad capabilities to simulate a wide range of general aeronautic conditions. General-purpose WTs can usually provide two generic facility tests: force and moment loads tests, and flow control and

separation tests. They can be characterized by the size of the test section and whether high or moderate Reynolds number (Rn)[11] capabilities are available.

Large tunnels are needed to handle the larger models of large vehicles or models that contain complex shapes under study. Smaller tunnels are needed in some circumstances to provide more cost-effective testing for smaller vehicles (such as general aviation or missiles).

Airflow behavior changes nonlinearly and unpredictably with changes in Rn. Thus, it is important to test the flow conditions at flight (or near-flight) Rn to ensure that the flows behave as expected and conditions such as undesired turbulence, separations, and buffeting do not occur. As a result, high-Rn facilities are needed. Larger vehicles such as transports fly at the highest Rn's, while fighters and unmanned combat aerial vehicles (UCAVs) operate at lower Rn's. Thus, transport aircraft have the highest Rn demands, while the tactical aircraft community's Rn needs are lower.

Unfortunately, testing in high-Rn facilities is expensive and can be much less productive than testing at low-Rn (atmospheric) facilities. Given the need to test as much of the flight envelope as possible, atmospheric facilities are also needed to produce the bulk of the test data.

---

[11] The Reynolds number is a nondimensional parameter describing the ratio of momentum forces to viscous forces in a fluid. The Mach number is a more familiar nondimensional parameter, describing the ratio of velocity to the sound speed in the fluid. When the flows around similarly shaped objects share the same nondimensional parameters, the topology of the flow for each will be identical (e.g., laminar and turbulent flow distribution, location of separation points, wake structure), and the same aerodynamic coefficients will apply (Batchelor, 1967). A wind tunnel experiment's results are only strictly applicable to flight conditions with the same Reynolds and Mach numbers, but it is very common to test only with matching Mach numbers, coming up short of the full-flight Reynolds number. Matching both parameters can be difficult because it requires changes in the working fluid such as higher pressure or lower temperature, or a different gas. Experience with similar airframes allows engineers to predict how vehicle aerodynamics will change between the test condition and flight condition, and often such extrapolations will prove adequate (particularly if the underlying flow structure changes slowly with Rn). However, it is not uncommon for a nuance of a newer design to invalidate predictions based on previous experience. Without testing at full-flight Reynolds number, there is *always* a risk of incorrect estimation.

For subsonic speeds, where costs are less intensive, a large high Rn and large atmospheric facility would satisfy most strategic needs. For transonic speeds, where testing costs start to become an issue, large- and medium-size facilities are warranted to help adjust for vehicle size and keep testing costs under control. Finally, for supersonic facilities, large, medium, and small facilities are needed for both high Rn and atmospheric testing to allow testers to adjust size and costs to vehicle size and need.

General-Purpose Hypersonic WTs

General-purpose hypersonic WTs provide force and moment load tests, as well as other tests, such as aerothermodynamic heating, flow over control surfaces, propulsion-airframe integration, and exhaust effects. Unlike conventional (subsonic through supersonic) WTs, general-purpose hypersonic facilities are perhaps best characterized by the hypersonic speed range they cover. Test capabilities should cover as much of a vehicle's speed range as possible. Therefore, we need low-, moderate-, and high–Mach number facilities.

Special-Purpose WTs

Special-purpose WTs are needed to meet especially demanding needs or when specialized test equipment (such as icing or propulsion simulation) is required. Specific categories of special-purpose test facilities are described below.

A very large, atmospheric, subsonic testing facility is needed to allow full-scale testing (often at near-flight Rn), especially for force and moment, control, and rotor performance in airflows for rotorcraft vehicles. Such vehicles and components are too complex to be scaled down in models and tested in normal-size, large facilities.

Very-high-Rn testing is needed given adverse Rn-scaling experiences on a variety of (especially larger) vehicles at the transonic level.

Propulsion simulation facilities at subsonic, transonic, and supersonic speeds are needed for all vehicle types to explore exhaust effects and propulsion-airframe integration.

Acoustics testing is a special need that is met by adding acoustic treatments to general-purpose subsonic facilities. Large-field acoustic measurements require a very large test section, while near-field acoustic measurements can be made in regular-size large facilities.

Aeroelasticity testing is needed at both subsonic and transonic speeds to understand and resolve such problems as fixed-wing flutter and buffet (transonic), dynamics, and divergence, as well as rotary-wing performance, loads, and stability.

Finally, special-purpose subsonic facilities are needed to test spin recovery, low-turbulence conditions, and icing, respectively. Transonic separation of weapon stores from vehicles is also needed for military vehicles.

Special-Purpose Hypersonic Propulsion Integration Test Facilities

Special-purpose hypersonic propulsion integration test facilities are needed to explore a number of different challenges to air-breathing hypersonic propulsion concepts. These include ramjet/scramjet, integrated engine, longer-duration, non-vitiated air, and high–Mach number testing.

General-Purpose Direct-Connect Propulsion Test Facilities

General-purpose direct-connect propulsion test facilities provide test capabilities for air-breathing jet engines where the input air is directed into the engine. The dominant difference between direct-connect facility needs is the size of the engine or component to be tested. The nation generally needs large, medium, and small direct-connect facilities. NASA research focuses on the small- and medium-size facilities, while the nation's large-engine testing needs are well met by AEDC.

Table 2.2 summarizes from our analyses the types of WT/PT facilities needed strategically for the nation to maintain its ability to conduct R&D across all aspects of aeronautics and propulsion and to ensure the capability to produce space vehicles.[12] With only a few exceptions for some of the smaller WT/PT facilities, each type of

---

[12] Details regarding the types of prediction tests conducted in each facility type are included in the companion technical report to this monograph (Antón et al., 2004[TR]).

Table 2.2
Test Facility Types Needed for National Strategic Reasons

Facility Type	Strategic National Need?
Subsonic WT	
General-Purpose	
Large, high Rn	Yes
Small, high Rn	No
Large, atmospheric	Yes
Small, atmospheric	No
Special-Purpose	
Propulsion simulation	Yes
Very large, atmospheric	Yes
Icing	Yes
Far-field acoustics	Yes
Near-field acoustics	Yes
Spin	Yes
Low turbulence	Yes
Dynamics	Yes
Transonic WT	
General-Purpose	
Large, high Rn	Yes
Medium, high Rn	Yes
Small, high Rn	No
Large, atmospheric	Yes
Medium, atmospheric	Yes
Small, atmospheric	No
Special-Purpose	
Very-high Rn	Yes
Store separation	Yes
Propulsion simulation	Yes
Dynamics/filter	Yes
Supersonic WT	
General-Purpose	
Large, high Rn	Yes
Medium, high Rn	Yes
Small, high Rn	Yes
Special-Purpose	
Propulsion simulation	Yes
Hypersonic WT	
General-Purpose	
Low Mach	Yes
Moderate Mach	Yes
High Mach	Yes
Special-Purpose	
Real gas effects	Yes

Table 2.2—Continued

Facility Type	Strategic National Need?
Hypersonic Propulsion Integration	
Ramjet/scramjet	Yes
Integrated engine test	Yes
Longer-duration testing	Yes
Non-vitiated	Yes
High Mach	Yes
Direct-Connect Propulsion	
Large engine and component	Yes
Medium engine and component	Yes
Small engine and component	Yes

facility is still needed to answer engineering RDT&E questions that cannot be answered computationally or efficiently through flight testing.

Summary

In summary, although applied aeronautics encompasses relatively mature science and engineering disciplines, there is still significant art and empirical testing involved in predicting and assessing the implications of the interactions between aeronautic vehicles and the environments through which they fly. This is, of course, especially true for complex new concepts that are not extensions of established systems with which engineers have extensive practical design and flight experience. But even improving the performance at the margin of well-established and refined designs—for example, commercial jet liners in areas such as reduced drag, fuel efficiency, emissions, noise, and safety (e.g., in adverse weather)—depends on WT/PT facility testing.

The fact that aeronautic R&D involves a fair amount of "cut and try" artistry makes the general and continuing need for aerodynamic testing clear, but it does not tell us about how valuable specific facilities might be. Designers are often surprised by what they find in testing their concepts despite decades of design experience and dramatic advances in CFD. Thus, while CFD has made inroads in reducing *some* empirical test simulation requirements, the technology will not replace the need for test facilities for the foreseeable future. Flight testing complements but does not replace WT/PT facilities because of the high costs and instrumentation limitations. The aeronautic engineering community does not have well-accepted handbooks of facility testing "best practices" or even rules of thumb from which to deduce testing requirements, nor is it possible from historical data to accurately predict returns on specific facility testing in terms of programmatic cost savings or risk reduction.[13]

The nation continues to need general-purpose R&D and T&E testing facilities across all speed regimes, as well as for specialty tests. These facilities advance aerospace research, facilitate vehicle design and development, and reduce design and performance risks in aeronautic vehicles.

---

[13] In theory, if one had access to historical data (including proprietary data from aeronautic manufacturers) from past failures and realistic estimates of the kind of testing program that could have prevented them, then one could calculate past returns on testing investments. Unfortunately, such data either does not exist or is not easily obtained because of proprietary concerns and a lack of documentation on tests considered but not implemented earlier in a program. This approach would also not address the life-long benefits that might have accrued from a more optimal design resulting from the knowledge gained from added testing (e.g., reduced drag resulting in fuel savings, increased payload capability, or added range). Moreover, the data are not necessarily predictive of future returns on investment, since the vehicle complexity and the physical domains challenged by new vehicle designs are not the same. A quantitative study of this question might be attempted to gain such a nonconclusive, historical quantitative support for the engineering premise that early and rigorous testing prevents future, expensive failures, but such an effort was beyond the scope of this study.

CHAPTER THREE

How Well Aligned Are NASA's Portfolio of Wind Tunnel and Propulsion Test Facilities with National Needs?

Given the strategic national needs for wind tunnel and propulsion test facilities described in Chapter Two, we next ask how well NASA's portfolio of existing facilities aligns with or satisfies those needs. NASA has 31 existing WT/PT facilities within the study's six categories. These facilities cover all flight regimes from subsonic to hypersonic, providing aerodynamic, propulsion, and specialized (e.g., icing, noise, and emissions) test capabilities. Although the facilities are complemented by non-NASA facilities (in DoD, abroad, and in the private sector), as well as by computational methods such as computational fluid dynamics and by flight tests, in this chapter we are concerned with how well NASA's portfolio of WT/PT facilities in isolation aligns with the strategic national needs described earlier.

Poor alignment could be of two types. First, there could be substantial *gaps* in coverage—that is, there would be needs that were not being met by any of NASA's WT/PT facilities. Second, there could be significant *overlaps*—that is, multiple NASA test facilities could align with one need.

Based on our research, we found that **nearly all existing NASA facilities serve one or more strategic need category (i.e., are primary facilities serving at least one national need) that is important to the nation's continuing ability to pursue aeronautic vehicles across NASA's roles of R&D, T&E, and strategic national interests. We found very few gaps in coverage.**

The remainder of this chapter provides support for these findings.

The Alignment of NASA's Portfolio of Wind Tunnels

NASA's 31 facilities align against the strategic needs for the six types of WT/PT facilities as shown in Table 3.1. The following six subsections discuss the alignment we find within those six categories. More detail on each of the 31 WTs can be found in the companion report to this monograph (Antón et al., 2004[TR]).

For each category, RAND staff reviewed the capabilities of each NASA WT/PT facility relative to the need categories listed in Table 3.1. This involved examining not only the primary capability of a facility but also its secondary capabilities. Engineering judgment was employed to determine the primary role of a facility with an eye to trying to serve the need categories with the least number of test facilities.

Subsonic WTs
Table 3.1 presents the alignment of NASA's eight subsonic WTs to serve strategic needs. In meeting national strategic needs across the general- or special-purpose categories, a single WT facility usually has a primary role in meeting such needs in the category and, perhaps, a secondary role in meeting needs in another category. All entries in bold represent the primary facility serving the need, while those non-bolded entries signify redundant facilities for that need. Superscript "a" in the table indicates that the facility is playing a secondary role in serving the needs in the category. For example, in the case of the Langley 14×22-Foot Atmospheric WT, the facility plays a primary role in serving the needs in the large, atmospheric category and a secondary role in serving the needs in the dynamics category.

Table 3.1
Alignment of NASA's Eight Subsonic WTs Against Strategic Needs

Facility Type	Existing NASA Facilities
General-Purpose	
Large, high Rn	**Ames 12-Foot High-Rn Pressure**
Large, atmospheric	**Langley 14×22-Foot Atmospheric**
	Langley 12-Foot Atmospheric Lab
Special-Purpose	
Propulsion simulation	**Glenn 9×15-Foot Propulsion**
	Ames 80×120-Foot and 40×80-Foot Atmospheric National Full-Scale Aerodynamics Complex[b]
Very large, atmospheric	**Ames 80×120-Foot and 40×80-Foot Atmospheric National Full-Scale Aerodynamics Complex**
Icing	**Glenn Icing Research Tunnel**
Far-field acoustics	**Ames 80×120-Foot and 40×80-Foot Atmospheric National Full-Scale Aerodynamics Complex[a]**
Near-field acoustics	**Glenn 9×15-Foot Atmospheric Propulsion[a]**
	Langley 14×22-Foot Atmospheric[a]
Spin	**Langley 20-Foot Vertical Spin Tunnel**
Low turbulence	**Langley Low-Turbulence Pressure Tunnel**
Dynamics	**Langley 14×22-Foot Atmospheric[a]**

[a]Secondary role.
[b]The National Full-Scale Aerodynamics Complex is speed-limited compared with the 9×15-Foot.

Looking across Table 3.1, we see that there is very good alignment, with few gaps and nearly all the WTs proving to be primary facilities. Only one WT—the Langley 12-Foot Atmospheric Lab—is shown to be redundant, since the Langley 14×22-Foot Atmospheric meets the need in this category.

Transonic WTs

Table 3.2 shows the alignment of NASA's five transonic WTs to serve strategic needs. Again, we see that almost all the facilities serve a national need and there are no gaps. The only redundant facility is the Langley 16-Foot Atmospheric Lab. For the large, atmospheric

Table 3.2
Alignment of NASA's Five Transonic WTs Against National Needs

Facility Type	Existing NASA Facilities
General-Purpose	
Large, high Rn	**Ames 11-Foot High-Rn**
Medium, high Rn	**Langley National Transonic Facility Very-High-Rn**[a]
Large, atmospheric	Langley 16-Foot Atmospheric
	Ames 11-Foot High-Rn[a]
Medium, atmospheric	**Langley National Transonic Facility Very-High-Rn**[a]
Special-Purpose	
Very-high Rn	**Langley National Transonic Facility Very-High-Rn**
Propulsion simulation	**Glenn 8×6-Foot Propulsion**
	Langley 16-Foot Atmospheric[a]
Dynamics/flutter	**Langley High-Rn Transonic Dynamics Tunnel**

[a]Secondary role.

general-purpose WT need, the Ames 11-Foot High-Rn Lab (in a secondary role) can meet the needs in the large, atmospheric category. For the propulsion simulation need where the Langley 16-Foot Atmospheric Lab is in its primary role, the Glenn 8×6-Foot Propulsion Lab can serve the needs. Additional details on these comparisons (including discussions of utilizations at these facilities and the technical competitiveness of them) can be found below and in the companion document (Antón et al., 2004[TR]).

Supersonic WTs

Table 3.3 shows the alignment of NASA's three supersonic WTs to serve strategic needs. In this category, all the NASA WTs serve national needs.

In the case of large, high-Rn capability, our analysis identifies a gap within NASA for serving this need. However, the DoD serves

Table 3.3
Alignment of NASA's Three Supersonic WTs Against National Needs

Facility Type	Existing NASA Facilities
General-Purpose	
Large, high Rn	(Gap within NASA)
Medium, high Rn	**Ames 9×7-Foot High-Rn**
Small, high Rn	**Langley 4-Foot High-Rn**
Special-Purpose	
Propulsion simulation	**Glenn 10×10-Foot Propulsion**

this national need well with the AEDC 16-Foot Supersonic Propulsion Wind Tunnel (16S). Given the low utilization for that facility, it is hard to argue that NASA needs a parallel capability in the large, high-Rn supersonic category.

Hypersonic WTs

Table 3.4 shows our results for NASA's three hypersonic WTs. As is true with the supersonic WTs, all these facilities serve national needs.

Once again, we find a gap within NASA—in this case, in the high-Mach area. Langley used to satisfy this gap with three facilities that have since been closed: the Langley 22-Inch Mach 20 Helium WT, the Langley 20-Inch Mach 17 Nitrogen WT, and the Langley 60-Inch Mach 18 Helium. Tests in the high-Mach speed regime are currently served by industry facilities—the Aero Systems Engineering (ASE) 20-Inch Channel 9 hypersonic WT, Veridian (former Calspan Corporation) 48-Inch and 96-Inch Shock Tubes, and the CUBRIC Large-Energy National Shock Tunnel I—as well as the AEDC hypervelocity Range/Track G and Hypervelocity Impact Range S1.

Table 3.4
Alignment of NASA's Three Hypersonic WTs Against National Needs

Facility Type	Existing NASA Facilities
General-Purpose	
Low Mach	**Langley 20-Inch Mach 6 Air**
Moderate Mach	**Langley 31-Inch Mach 10 Air**
High Mach	(Gap within NASA)
Special-Purpose	
Real gas effects	**Langley 20-Inch Mach 6 Tetraflouromethane**

Hypersonic Propulsion Integration Test Facilities

Table 3.5 shows the alignment of NASA's nine hypersonic propulsion integration test facilities to serve strategic needs. Once again, we see that all the facilities serve national needs; however, two of these facilities—Ames 16-Inch Shock and Ames Direct-Connect—have been mothballed.[1]

In addition, one gap is identified, in this case, within the high-Mach area. Tests in the high-Mach regime are currently served by industry facilities: the Hypersonic Pulse (HYPULSE) facility operated by GASL for NASA, Aero Systems Engineering 20-Inch Channel 9 Hypersonic Wind Tunnel, Veridian (former Calspan Corporation) 48-Inch and 96-Inch Shock Tubes, or CUBRIC Large-Energy National Shock Tunnel I.

---

[1] *Mothballing* involves temporarily closing a facility while preserving it for future use. Mothballed facilities generally involve minimal maintenance and could be brought back on line (fully functional) in, say, two months or less. While mothballing an important facility is preferred to abandonment, it usually involves the loss of the workforce expertise required to safely and effectively operate the facility. Thus, mothballing is not a simple solution to dealing with long periods of low utilization, and it puts facilities at risk.

Table 3.5
Alignment of NASA's Nine Hypersonic Propulsion Integration Test Facilities Against National Needs

Facility Type	Existing NASA Facilities
Ramjet/scramjet suite	Langley 8-Foot High-Temperature Tunnel
	Langley Arc-Heated Scramjet
	Langley Combustion Scramjet
	Langley Supersonic Combustion
	Langley 15-Inch Mach 6 High-Temperature Tunnel
	NASA/GASL HYPULSE
Integrated engine tests	Ames 16-Inch Shock[b]
Longer-duration testing	Ames Direct-Connect[b]
	Glenn Propulsion Simulation Lab Cell 4[a]
Non-vitiated	Glenn Non-Vitiated Hypersonic Tunnel Facility
High Mach	(Gap within NASA)

[a]Secondary role.
[b]Mothballed.

Direct-Connect Propulsion Test Facilities

Table 3.6 shows the alignment of NASA's three direct-connect propulsion test facilities to serve strategic needs. As shown, all these facilities serve national needs.

Table 3.6
Alignment of NASA's Three Direct-Connect Propulsion Test Facilities Against National Needs

Facility Type	Existing NASA Facilities
Large engine and component	(Gap within NASA)
Medium engine and component	Glenn Propulsion Simulation Lab Cell 3
	Glenn Propulsion Simulation Lab Cell 4
Small engine and component	Glenn Engine Components Research Lab Cell 2b

Once again, we identified one gap within NASA: large engine and component. However, this national need is well served by the world's largest direct-connect propulsion test facilities: the AEDC Aeropropulsion Systems Test Facility Propulsion Development Test cells C1 and C2.

Summary

In summary, these facilities are generally consistent with the testing needs of NASA's research programs, as well as with those of the broader national R&D programs. Currently, redundancy is minimal across NASA. Facility closures in the past decade have eliminated almost a third of NASA's test facilities in the categories under review in this study. In nearly all test categories, NASA has a single facility that serves the general- or special-purpose testing needs, although some primary facilities also provide secondary capabilities in other test categories. In other words, nearly all facilities that plays a secondary role in one category play a unique primary role in a different category; therefore, these facilities could not be eliminated simply on the basis of playing that secondary role. Put another way, while there is certainly significant unused capacity in most NASA facilities, *testing could not be consolidated in a significantly smaller set of NASA facilities to reduce costs*. As such, 29 of the 31 WT/PT facilities are unique in serving at least one national need (within NASA). Thus, the current type of test complex *within NASA* is both responsive to user needs and is mostly "right sized" to the range of national aeronautic engineering needs.

Two redundant WTs were found: (1) the Langley 12-Foot Subsonic Atmospheric Wind Tunnel Lab, which is redundant to the Langley 14×22-Foot Subsonic Atmospheric WT, and (2) the Langley 16-Foot Transonic Atmospheric Wind Tunnel (16TT), which is generally redundant to the Ames 11-Foot Transonic High-Rn and Langley National Transonic Facility WTs run in low-Rn conditions.

There are gaps in NASA's ability to serve all national needs. In most of those cases, DoD or commercial facilities serve the national needs.

Finally, some of NASA's facilities that serve national needs have been or are in the process of being mothballed. While mothballing an important, complex facility is preferred to abandonment, it involves the loss of workforce expertise required to safely and effectively operate the facility. Thus, mothballing is not a simple solution to dealing with long periods of low utilization and puts facilities at risk.

CHAPTER FOUR

What Is the Condition of NASA's Portfolio of Wind Tunnel and Propulsion Test Facilities?

Given that nearly all of NASA's existing wind tunnel and propulsion test facilities can be mapped back to important national needs, the next question has to do with what condition those facilities are in (i.e., how healthy they are). Even though a test facility may be the only NASA facility serving a national need, it may have serious issues about how *technically competitive* it is (in terms of serving state-of-the-art requirements) and how well it is *utilized*. Another dimension in determining the health or condition of NASA's portfolio of WT/PT facilities is the *financial health* of any given facility. Facilities that are technically competitive but poorly utilized tend to be in poor financial health.

Based on our research, we find that **84 percent of NASA's WT/PT facilities are technically competitive and effective with state-of-the-art requirements, although there is room for improvement.** When we further consider the question of utilization, we find that **a little more than two-thirds (68 percent) are well utilized.** Issues of financial health are driven by the utilization these facilities have and the financing arrangements under which the facilities operate. Given this, we find that **about a third (32 percent) of the facilities have severe financial health issues. For such facilities, NASA's drive for full-cost recovery (FCR) can lead to financial collapse.**

We discuss these findings in more detail below.

Technical Competitiveness of NASA's WT/PT Facilities

To determine the technical competitiveness of NASA's WT/PT facilities, we relied on a number of internal and external sources across NASA, the DoD, and industry. This assessment was based on expert assessment of the facility's technical capabilities, in-depth surveys of the user base and design community, and reviews of published research and assessments. When published data were not available, assessments and surveys of experts based on testing experience in the RDT&E communities were employed. More details on the approach and findings can be found in the companion document (Antón et al., 2004[TR]).

Based on these analyses, we rated each of NASA's 31 WT/PT facilities in terms of its technical competitiveness. Table 4.1 summarizes these facility ratings, with the last column summarizing whether or not the facility is technically competitive. Facilities were identified as being technically competitive ("strong") when they had good, robust technical capabilities that will be required in RDT&E of current and next-generation aerospace vehicles. Technically "weak" facilities have one or more technical limitation, as noted in the assessment column, that significantly affects their ability to meet needs.

Utilization of NASA's Facilities

To determine how well utilized NASA's WT/PT facilities are, we relied on historical utilization data from NASA that plots utilization in terms of UOH. Utilization was broken down in each fiscal year by the sector that paid for the test: NASA (directly for a NASA research program), a military program (whether directly for a DoD organization or for a military contractor), commercial industry for commercial research or vehicle production, or a cooperative program in which NASA paid for the test costs. In some cases, quantitative data on past utilization were not available; for these cases, summary utilization assessments from NASA were employed.

Table 4.1
Summary of Assessments of Technical Competitiveness

Existing NASA Facilities	Assessment of Technical Competitiveness	Synopsis
Subsonic WT		
Ames 12-Foot High-Rn Pressure	Only high-Rn subsonic facility in the United States, but has some undesirable features and limitations that render it unacceptable for some user communities (both commercial transport and tactical aircraft development); technical capabilities are fine for other communities. Two foreign facilities are more advanced in many respects.	Weak
Ames 80×120-Foot and 40×80-Foot Atmospheric National Full-Scale Aerodynamics Complex	Strong enabling capabilities for rotorcraft and other low-speed flight Rn capabilities as well as airframe noise, inlet instrumentation, and base-flow analysis. Immense size allows full-scale testing of smaller vehicles.	Strong
Glenn 9×15-Foot Propulsion	Excellent propulsion simulation capabilities. Supports noise-reduction research. Low Rn and no force testing capability.	Strong
Glenn Icing Research Tunnel	Best icing research facility in the world. Excellent research staff.	Strong
Langley 12-Foot Atmospheric Lab	Has very-low-Rn capabilities, poor user advocacy outside Langley, and is viewed as unacceptable by important industry segments. Very low cost to operate.	Weak
Langley 14×22-Foot Atmospheric	Good quality atmospheric facility. Low cost to operate.	Strong
Langley 20-Foot Vertical Spin Tunnel	Best capability available for spin studies.	Strong
Langley Low-Turbulence Pressure Tunnel	Unique high-Rn research capabilities.	Strong
Transonic WT		
Ames 11-Foot High-Rn	One of top two U.S. high-Rn facilities. Good Mach number range. No propulsion simulation capability.	Strong
Glenn Transonic 8×6-Foot Propulsion	Excellent propulsion simulation capability. Excellent Mach number range. Low Rn and no force measurement capability.	Strong
Langley High-Rn Transonic Dynamics Tunnel	Unique and capable facility for transonic dynamics. Limit Mach number range. Non-air fluid could introduce data interpretation questions, but has not been a problem to date.	Strong

Table 4.1—Continued

Existing NASA Facilities	Assessment of Technical Competitiveness	Synopsis
Transonic WT (cont.)		
Langley National Transonic Facility Very-High Rn	Excellent very-high-Rn research capability. Low productivity for cryogenic operation. Model dynamics limitations at cryogenic temperatures. Being upgraded to operate in air for high-productivity, low-Rn tests.	Strong
Langley Transonic 16-Foot Atmospheric	Has low Rn capability, poor flow quality outside the center 4 feet of airflow, and a low maximum Mach number. Inferior to Ames 11-Foot, Langley National Transonic Facility, and AEDC 16-Foot Transonic facilities.	Weak
Supersonic WT		
Ames Supersonic 9×7-Foot High-Rn	High Rn. Compatible with Ames 11-Foot. Low maximum Mach number of 2.5 a concern. No propulsion simulation capability.	Strong
Glenn Supersonic 10×10-Foot Propulsion	Excellent propulsion simulation capability but low Rn and no force testing capabilities. Maximum speed of Mach 3.5 a plus, but minimum speed of 2.0 is a limitation.	Strong
Langley Supersonic 4-Foot High-Rn	Excellent Mach number range. Continuous flow a plus compared with other 4-foot WTs. Very cost-effective for small vehicles compared with Ames 9×7-Foot and AEDC 16-Foot supersonic WTs.	Strong
Hypersonic WT		
Langley Hypersonic 20-Inch Mach 6 Air	Best understood as a suite of capabilities offering a good choice for preliminary design and some real gas effects: good for research; aeroheating capability; some store separation capabilities; lower cost compared with AEDC facilities makes these facilities good places for preliminary design with maturing projects moving to AEDC.	Strong
Langley Hypersonic 20-Inch Mach 6 Tetra-flouromethane		Strong
Langley Hypersonic 31-Inch Mach 10 Air		Strong
Hypersonic Propulsion Integration		
Ames Hypersonic Propulsion Integration 16-Inch Shock	Strong for integrated engine tests for benchmarking. Good for research. Used extensively in the past, but are mothballed and in questionable condition.	Weak
Ames Hypersonic Propulsion Integration Direct-Connect	Strong capability for longer-duration testing at velocities greater than 7,000 feet per second for modest cost. Good for providing benchmark data to anchor test data needed for modeling. Used extensively in the past, but are mothballed and in questionable condition.	Weak

Table 4.1—Continued

Existing NASA Facilities	Assessment of Technical Competitiveness	Synopsis
Hypersonic Propulsion Integration (cont.)		
Glenn Non-Vitiated Hypersonic Tunnel Facility for Propulsion Integration	Unique non-vitiated airflow capability. Size and speed limited.	Strong
Langley Hypersonic Propulsion Integration 15-Inch Mach 6 High-Temperature Tunnel	The five facilities at Langley plus HYPULSE are best understood as a suite of capabilities covering the operational envelope for scramjet propulsion R&D with generally subscale models. Their biggest drawbacks are size and test duration. Mach number and enthalpy ranges are adequate. Most have the disadvantage of vitiating the airflow during heating.	Strong
Langley Hypersonic Propulsion Integration 8-Foot High-Temperature Tunnel		Strong
Langley Hypersonic Propulsion Integration Arc-Heated Scramjet		Strong
Langley Hypersonic Propulsion Integration Combustion Scramjet		Strong
Langley Hypersonic Propulsion Integration Supersonic Combustion		Strong
NASA/GASL HYPULSE Propulsion Integration		Strong
Direct-Connect Propulsion		
Glenn Propulsion Simulation Lab Cell 3	Two test cells in a unified facility. Strong capability for testing small engines and components for business jets, regional jets, fighters, etc. Similar to capabilities at AEDC.	Strong
Glenn Propulsion Simulation Lab Cell 4		Strong
Glenn Engine Components Research Lab Cell 2b	Strong facility for small turbine engine and components testing.	Strong

To get some sense of near-term future utilization, we relied on the results of our survey of user representatives in December 2002 in which we asked users to complete a spreadsheet indicating anticipated UOH for WT facilities and EOH for PT facilities. We found these data to be unreliable beyond a few months into the future, since programs do not have quantitative testing plans more than a few months to a few years into the future and revisions are common. NASA test facility managers confirmed this difficulty.

As a result, utilization data do not serve as a practical metric for the long-term need of a facility but can be useful to gain insights into other aspects of facility health (e.g., for current financial health as shown below). The strategic assessments used in Chapter Two are better measures for facility need.

Based on our analysis of these data, we assessed each of NASA's 31 WT/PT facilities in terms of its utilization. Table 4.2 summarizes our assessments of those facilities and includes in the final column an assessment of whether the facility is well utilized ("good") or not ("poor"). Utilization ratings are summary assessments of quantitative data (when available) and qualitative data, including the overall levels of use relative to current staff capacity and near-term demand. Facilities in financial difficulty due to low utilization were obvious candidates for "poor" ratings. For details, see Antón et al. (2004[TR]).

Table 4.2
Summary of Assessments of Utilization

Existing NASA Facilities	Assessment of Current Usage	Synopsis
Subsonic WT		
Ames 12-Foot High-Rn Pressure	Commercial aircraft usage never recovered from the extended shutdown time when users established preference for a foreign facility. Good advocacy from certain sectors (e.g., space access and unmanned aerial vehicles [UAVs]) but no current usage. (quantitative data)	Poor
Ames 80×120-Foot and 40×80-Foot Atmospheric National Full-Scale Aerodynamics Complex	Sparse utilization once NASA ended its rotorcraft research program. Noted parachute tests for Mars lander. Vocal advocacy. (quantitative data)	Poor
Glenn 9×15-Foot Propulsion	Consistently near deliverable capacity. (quantitative data)	Good
Glenn Icing Research Tunnel	Consistently at deliverable capacity—very-high-demand facility. (quantitative data)	Good
Langley 12-Foot Atmospheric Lab	Consistently low past and projected utilization. As a lab, this facility is not broadly used by the users of the large, state-of-the-art WT facilities. (quantitative data)	Poor

Table 4.2—Continued

Existing NASA Facilities	Assessment of Current Usage	Synopsis
Subsonic WT (cont.)		
Langley 14×22-Foot Atmospheric	Consistently at or above deliverable capacity (quantitative data)	Good
Langley 20-Foot Vertical Spin Tunnel	Current usage appears sparse and dominated by the DoD, but no quantitative data available. Has been used by non-DoD programs in the past. (qualitative data)	Poor
Langley Low-Turbulence Pressure Tunnel	Current usage may be OK based on excellent capabilities, but quantitative data was available. (qualitative data)	Good
Transonic WT		
Ames 11-Foot High-Rn	Utilization has not been as high as it was before the extended shut down for modernization, but current usage has been enough to keep the facility and the Ames complex open. (quantitative data)	Good
Glenn Transonic 8×6-Foot Propulsion	Consistently near deliverable capacity. (quantitative data)	Good
Langley High-Rn Transonic Dynamics Tunnel	Current usage appears sparse and dominated by the DoD with some NASA research, but no quantitative data available. (qualitative data)	Poor
Langley National Transonic Facility Very-High-Rn	Utilization is and has been very strong—at or near deliverable capacity. New air-only capability will likely increase demand. (quantitative data)	Good
Langley 16TT	Utilization is and has been very strong—at or near deliverable capacity. (quantitative data)	Good
Supersonic WT		
Ames Supersonic 9×7-Foot High-Rn	Current and historical utilization has been moderate to poor due to current lack of supersonic programs. Usage may pick up from supersonic business jet concepts. (quantitative data)	Poor
Glenn Supersonic 10×10-Foot Propulsion	Current usage is significantly lower than historical levels, although management has adjusted staffing and deliverable capacity to accommodate. (quantitative data)	Poor
Langley Supersonic 4-Foot High-Rn	Consistently at or above deliverable capacity. (quantitative data)	Good

Table 4.2—Continued

Existing NASA Facilities	Assessment of Current Usage	Synopsis
Hypersonic WT		
Langley Hypersonic 20-Inch Mach 6 Air	Low cost and good capabilities have made these three WTs attractive for preliminary design stages of hypersonic research programs. The current upswing in hypersonic research has made usage relatively stable, but exact levels may fluctuate or be low at times. (qualitative data)	Good
Langley Hypersonic 20-Inch Mach 6 Tetra-flouromethane		Good
Langley Hypersonic 31-Inch Mach 10 Air		Good
Hypersonic Propulsion Integration		
Ames Hypersonic Propulsion Integration 16-Inch Shock	Facility currently in mothball; therefore no usage.	Poor
Ames Hypersonic Propulsion Integration Direct-Connect	Facility currently in mothball; therefore no usage.	Poor
Glenn Non-Vitiated Hypersonic Tunnel Facility for Propulsion Integration	Current utilization is poor, but demand may rise quickly due to non-vitiated capability. (qualitative data)	Poor
Langley Hypersonic Propulsion Integration 15-Inch Mach 6 High-Temperature Tunnel	Low cost and good capabilities have made the suite of five facilities at Langley plus HYPULSE attractive for preliminary design stages of hypersonic research programs. The current upswing in hypersonic research has made usage relatively stable, but exact levels may fluctuate or be low at times. The High-Temperature Tunnel in particular had poor utilization for years but is now cited as a critical facility that will be used by current hypersonic research programs. (qualitative data)	Good
Langley Hypersonic Propulsion Integration 8-Foot High-Temperature Tunnel		Good
Langley Hypersonic Propulsion Integration Arc-Heated Scramjet		Good
Langley Hypersonic Propulsion Integration Combustion Scramjet		Good
Langley Hypersonic Propulsion Integration Supersonic Combustion		Good
NASA/GASL HYPULSE Propulsion Integration		Good
Direct-Connect Propulsion		
Glenn Propulsion Simulation Lab Cell 3	Usage has been moderately good for two decades, although below deliverable capacity. (quantitative data)	Good

Table 4.2—Continued

Existing NASA Facilities	Assessment of Current Usage	Synopsis
Direct-Connect Propulsion (cont.)		
Glenn Propulsion Simulation Lab Cell 4	Usage has been moderately good for two decades, although below deliverable capacity. Despite operational capability to Mach 6, Propulsion Simulation Lab–4 is not often used for hypersonic propulsion testing. (quantitative data)	Good
Glenn Engine Components Research Lab Cell 2b	Current usage OK, but no quantitative data available. Usage is dominated by NASA and Army programs (e.g., rotorcraft engines) with some industry advocacy. (qualitative data)	Good

Assessing NASA's Portfolio of WT/PT Facilities in Terms of Technical Competitiveness and Utilization

The matrix in Table 4.3 groups NASA's WT/PT facilities according to their technical competitiveness and current utilization, combining the assessments from the analysis discussed above.

The upper right-hand quadrant represents those facilities ranked both as strongly technically competitive and with good usage (i.e., they serve at least one national need with good technical capabilities and have good current utilization). Twenty facilities fall into this category. Technically competitive facilities with poor usage (the lower right-hand quadrant) are competitive with current and future state-of-the-art requirements but currently have low utilization because of the cyclic nature of usage. Six of the facilities fall into this category. All told, 26 facilities (84 percent) rank as highly technically competitive, while 21 (about 68 percent) rank as well utilized.

Facilities that are weakly competitive (technically) with good usage (the upper-left quadrant) are not competitive with state-of-the-art technical requirements despite current high utilization resulting from one of many reasons (e.g., low cost, ready availability, entrenched use, familiarity, or existing databases). Only one of the WT/PT facilities falls into this category. Facilities that are weakly competitive (technically) with poor usage (the lower-left quadrant) are not only weakly competitive technically but are also sparsely used. The remaining four WT/PT facilities fall into this category.

Another way to look at the results in Table 4.3 is by WT/PT categories. Table 4.4 divides the facilities by category in terms of whether they are technically competitive, well utilized, and both technically competitive and well utilized. All the hypersonic and direct-connect propulsion test facilities are both technically competitive and well utilized,[1] whereas the supersonic WTs, while all technically competitive, are considerably underutilized at this time.[2]

Room for Improvement in Technical Competitiveness

Despite being generally consistent with NASA's current testing needs, there remain areas where today's facilities need upgrades or replacement to make them more competitive with state-of-the-art requirements.

There are many facilities that operate at Rn's well below flight conditions. Experience to date has proven that both empirical tests in high-Rn facilities and actual flight experience often reveal surprises not observable in low-Rn WT tests. Rn effects are nonlinear, and engineers cannot reliably predict when a small modification to a robust design will trip an airflow threshold and cause undesirable behavior. Recent examples include comparisons of low- and high-Rn test results,[3] the underestimation of the cruise Mach number for the Boeing 777 (preventing the missed opportunity to trade speed for

---

[1] These WTs are well utilized as a result of the recent upswing in hypersonics programs and continuing refinement of jet engines.

[2] Supersonic WT usage is currently very low as a result of the current lack of commercial jet aircraft production and the use of extrapolation models for transonic WT data for military supersonic aircraft. However, commercial interest in supersonic business jets and the eventual expanded performance of military supersonic aircraft were cited in our surveys as examples of future need for supersonic WTs.

[3] See, for example, Curtin et al. (2002).

Table 4.3
Competitiveness and Current Usage of NASA's Wind Tunnel Facilities

Current Usage		Weakly Technically Competitive	Strongly Technically Competitive
Good		**1 out of 31 Facilities** • Langley Transonic 16-Foot Atmospheric	**20 out of 31 Facilities** • Glenn Subsonic Icing Research Tunnel • Glenn Subsonic 9×15-Foot Propulsion • Langley Subsonic 14×22-Foot Atmospheric • Langley Subsonic Low-Turbulence Pressure Tunnel • Ames Transonic 11-Foot High-Rn • Glenn Transonic 8×6-Foot Propulsion • Langley National Transonic Facility Very-High-Rn • Langley Supersonic 4-Foot High-Rn • Langley Hypersonic 20-Inch Mach 6 Air • Langley Hypersonic 31-Inch Mach 10 Air • Langley Hypersonic 20-Inch Mach 6 Tetraflouromethane • Langley Hypersonic Propulsion Integration 8-Foot High-Temperature Tunnel • Langley Hypersonic Propulsion Integration Arc-Heated Scramjet • Langley Hypersonic Propulsion Integration Combustion Scramjet • Langley Hypersonic Propulsion Integration Supersonic Combustion • Langley Hypersonic Propulsion Integration 15-Inch Mach 6 High-Temperature Tunnel • NASA/GASL HYPULSE Propulsion Integration • Glenn Propulsion Simulation Lab Cell 3 • Glenn Propulsion Simulation Lab Cell 4 • Glenn Engine Components Research Lab Cell 2b
Poor		**4 out of 31 Facilities** • Ames Subsonic 12-Foot High-Rn Pressure[a] • Langley Subsonic 12-Foot Atmospheric Lab • Ames Hypersonic Propulsion Integration 16-Inch Shock[a] • Ames Hypersonic Propulsion Integration Direct-Connect[a]	**6 out of 31 Facilities** • Langley Subsonic 20-Foot Vertical Spin Tunnel • Ames Subsonic 80×120-Foot and 40×80-Foot Atmospheric National Full-Scale Aerodynamics Complex • Langley High-Rn Transonic Dynamics Tunnel • Ames Supersonic 9×7-Foot High-Rn • Glenn Supersonic 10×10-Foot Propulsion • Glenn Non-Vitiated Hypersonic Tunnel Facility for Propulsion Integration

Technically Competitive with State-of-the-Art Needs

[a]Mothballed.

Table 4.4
Technical Competitiveness and Utilization by Test Facility Category

Category	Facilities (number)	Technically Competitive (%)	Well Utilized (%)	Technically Competitive and Well Utilized (%)
Subsonic WT	8	75	50	50
Transonic WT	5	80	80	60
Supersonic WT	3	100	33	33
Hypersonic WT	3	100	100	100
Hypersonic Propulsion Integration	9	78	66	66
Direct-Connect Propulsion	3	100	100	100
Overall	31	84[a]	68[a]	65[a]

[a]Percentages are based on the whole set of 31 facilities (e.g., 26/31 = 84%).

range or other trade-offs) and the cruise drag predictions of the C-141 and C-5 (resulting in significant fuel cost increases).[4] The only reliable solution is to test at higher-Rn WTs, but this typically requires more expensive, less productive, and sometimes larger facilities (particularly when test models cannot be effectively configured for smaller, high-Rn pressure tunnels). These high-Rn facilities are expensive, to both build and operate. These scaling issues are the single most critical shortfall, and one that may require significant investments to address.

Needs for high-productivity, high-Rn subsonic and transonic WTs have been voiced in the past—more recently in the National Facility Study[5] and recent Air Force testimony.[6] The users surveyed had difficulty supporting these needs—not on technical research or

---

[4] See Wahls (2001) and Crook (2002), as well as the NASA/DoD Flight Prediction Workshop in November 2002.

[5] *National Facilities Study* (1994).

[6] See Lyles (2002).

programmatic risk-reduction needs, but because the tight fiscal constraints that prevented the acquisition of these expensive facilities ($2–3 billion) in the late 1990s remain today in their programs. The nation's premier high-Rn facilities—the Ames 12-Foot High-Rn Pressure Wind Tunnel, the Langley Very-High-Rn National Transonic Facility, the Ames 11-Foot High-Rn Transonic WT, and the AEDC 16-Foot Transonic and 16-Foot Supersonic High-Rn WTs—are quite capable, but maintenance and modernization investments are not keeping up with the state of the art. For example, the Ames 12-Foot served the nation well in past decades but is now inferior to the QinetiQ 5-Metre in the United Kingdom in some aspects. Also, the National Transonic Facility is a generation behind the European Transonic Windtunnel.

The study identified other areas where future investments may be warranted. There is a significant, $128 million backlog of maintenance and repair (BMAR) that should be eliminated. This BMAR is approximately equivalent in dollars to the annual operating budget for NASA's facilities, Also, various upgrades to existing facilities are needed to address identified challenges and to regain lost capabilities. The most readily identifiable major investment need from our survey of users is associated with the hypersonic vehicle programs. Serious research challenges in hypersonic air-breathing propulsion (e.g., vitiated/non-vitiated[7] issues in hypersonic propulsion facilities) may require new facilities and test approaches for breakthroughs to occur. This will require research in test techniques to understand how to address these testing needs and ultimately to look at the viability of hypersonic propulsion concepts being explored.

---

[7] In hypersonics testing, the test gases representing airflow must be preheated before undergoing expansion to achieve hypersonic velocities at the test section. When combustion is used to heat the gases, it vitiates (affects the quality of) the gas mixture by changing the mixture and introducing combustion by-products. Air-breathing hypersonic propulsion concepts such as scramjets that rely on atmospheric oxygen for combustion may be affected by such vitiation, but the vitiated effects remain a research question that may need to be resolved before such propulsion concepts can be made practical. Non-vitiated heating does not change the quality of the test flow.

Assessing the Financial Health of NASA's Portfolio

The financial health of NASA's portfolio is another component in determining WT/PT facility health ratings. In this case, because of NASA's use of FCR for its facilities (discussed below), financial health is correlated with how well or poorly utilized the facilities are. In terms of Table 4.3 above, the bottom half of the table represents the WT/PT facilities that are the most underutilized; in this case, this represents a little less than a third of the facilities. However, as the "well utilized" column in Table 4.4 indicates, most categories of WT/PT facilities are well utilized. All the hypersonic and direct-connect propulsion facilities are well utilized, and the subsonic and transonic WTs are mostly well utilized. Supersonic facilities are generally in the worst shape.

How NASA Finances Its WT/PT Facility Operations

Abstractly, a research center at NASA can fund some test facility costs directly and expects users to fund others. Users pay their share of costs through a per-hour transfer price for testing at facilities and may incur other costs through center-wide taxing mechanisms. Under full-cost recovery *pricing*, users pay for all annual costs through transfer prices. Alternatively, full costs could be recovered through a mix of such transfer prices and separate program taxing mechanisms.[8] Eventually, each center compensates for differences between realized annual costs and revenues in some way.

NASA centers generally use forms of average-cost pricing[9] to determine the transfer prices for tests.[10] The granularity of the pricing varies by the actual accounting system used, but in general, each cen-

---

[8] According to NASA's *Full Cost Initiative Agencywide Implementation Guide* (1999), "Any differences between full cost recovery and full cost accounting must be budgeted against a project or included in the G&A pool" (paragraph 3.1.2.2, p. 27).

[9] In average-cost pricing, the price is set equal to the average total cost of production rather than, say, the marginal cost (Coutts, 1998). The average total cost is the total (full) cost per unit of output, found by dividing total cost by the quantity of output (Rutherford, 1992).

[10] These are "transfer" prices because most users of NASA facilities are NASA or other government programs; the prices are paid through intergovernmental fund transfers.

ter charges users a test price that spreads the costs that need to be recovered through pricing over the estimated number of test hours for the year. In its simplest form, the transfer price can be defined as follows:

$$\text{Transfer Price} = \frac{(\text{pool of annual costs that the center expects users to pay})}{(\text{expected hours of facility use for the year})}$$

Table 4.5 summarizes the current funding and cost pressures at the three primary NASA centers involved in aerodynamic and propulsion testing. Figure 4.1 shows the funding source(s) in FY2002 for the three centers' test facilities.

At Ames, the primary source of test funding is external sources outside NASA (i.e., military and commercial sectors), since the NASA rotorcraft program funding was eliminated in FY2003. At Langley, local NASA programs dominate test facility funding. At

Table 4.5
Funding and Cost Pressures at the Three NASA Centers

	Centers (focus)		
	Ames (IT)	**Glenn (propulsion)**	**Langley (aerodynamics)**
Primary source(s) of funding	Non-NASA users[a]	NASA programs and non-NASA users	NASA programs
Source of test complex shared support	Vanishing center support	Center support	Tax on programs (relatively larger business base)
Average-cost transfer price	Volatile: Small user base is the only source of funds	Volatility limited: Taxes on all center programs cover most overhead, including maintenance	Volatility limited: Unutilized capacity is covered by program user tax base, and utilization variance is limited by large user base

[a]NASA rotorcraft program funding was a major source for the National Full-Scale Aerodynamics Complex until recently.

Glenn, both internal NASA programs and external users are primary test facility funding sources.

Mechanisms of non-price-related FCR shared support continue for the test complexes at Langley and Glenn, but Ames is eliminating center support because of its focus on other missions. The average cost transfer price at Ames is volatile, since Ames has a small user base, annual variances in each user's utilizations, and few WT/PT facilities across which staff could be shared to help offset those variances.

Figure 4.1
Internal or External Sourcing of FY2002 Operating Funds for WT/PT Facilities at Ames, Glenn, and Langley[11]

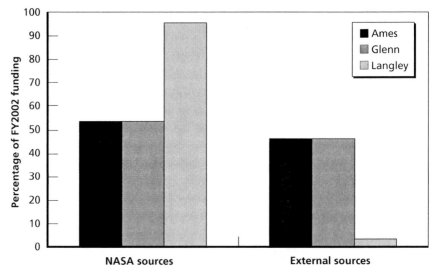

SOURCES: NASA Cost Workshop, January 22, 2003.
RAND MG178-4.1

---

[11] With the elimination of the NASA rotorcraft program and center shared support at Ames, external sources are becoming the dominant funding source of the center's test facilities.

All costs for the Langley facilities are recovered from the users of those facilities through test prices and tax mechanisms. Most small facilities are billed directly to the principal benefiting NASA program. Average-cost transfer prices for testing at Langley are expected to cover most facility costs and depend on the expected hours of facility use for the year. Taxes on the NASA program user base for each large Langley facility provide funding to cover any remaining unutilized capacity in each large facility, helping to stabilize the average-cost transfer prices. Also, the large size of those user bases dampens the utilization volatility to some extent.[12]

Likewise, all costs for the Glenn facilities are recovered from the users of those facilities through test prices and tax mechanisms. The transfer prices for testing at Glenn have been stable because users (NASA and external) pay only direct costs incurred for their test. This includes a nominal fixed weekly surcharge that covers a small percentage of the fixed overhead costs of keeping the facility open and maintained. The fixed overhead costs have primarily been borne by a combination of the center's Aeronautics Base Research and Technology Program and by the center's other aeronautics and space program offices through the Center-wide Program Support tax. However, the implementation of full-cost accounting and financing of the Glenn ground test facilities is in progress. The practical effect on customer costs and Glenn's ability to maintain and operate its R&D test facilities could not be defined by Glenn at this time.[13]

Funding Options

Given this picture of funding at the centers, we considered three basic ways NASA could recover the full annual costs of a facility:

1. Users fund the full annual costs using a transfer price based on average cost. A center funds nothing.

---

[12] Personal communications with Langley facility management.

[13] Personal communications with Glenn facility management.

2. Each center funds the full annual cost of a facility. Users fund nothing.
3. Each center splits the full annual funding of a facility with its users.

When Users Fund the Full Annual Costs Using a Transfer Price Based on Average Cost. Setting the transfer price to cover all costs can discourage use and endanger strategic facilities. This approach indeed gives users more information about the full costs for conducting their tests at a facility. If this cost is too high, users can respond by seeking an alternative source of services. Alternatively, users may avoid needed testing (Bergquist et al., 1972). What is important to note is that users (e.g., specific programs) do not have an explicit responsibility to maintain test capabilities. Thus, relying on users to pay the full costs of the test facility infrastructure is problematic at best.

This approach would be a good option if the alternatives are a better value over the long term and strategically important resources are retained. Unfortunately, this approach leads to poor outcomes if NASA is a better long-term value but low near-term utilizations and resulting higher near-term prices mask the long-term value. This approach is also bad when the remaining users cannot afford the costs to keep open strategic facilities needed in the long term.

If actual and expected revenues differ at the end of the year, centers must cover the difference. Federal financial management practices limit the centers' ability to manage large differences effectively over time. Effective use of such fund transfers requires planning coordination between users and centers. Centers bear the brunt of poor coordination.

When Each Center Funds the Full Annual Cost of a Facility. Setting the transfer price at zero can encourage overuse and cause limits on the availability of funding. This approach closes one channel of information to users about the costs they are imposing on the infrastructure. Users may respond by increasing their use of NASA facilities. Increased use is good if it increases access to high-quality NASA facilities that are key to advancing the state of the art in aeronautics. However, increased use is bad if effective subsidies mask NASA's lower value (compared with alternate or even proposals for new, more productive facilities) over long term. There may also be insufficient motivation to tailor capabilities to a user's real requirements (Bergquist et al., 1972).

This approach simplifies budgeting and funds management in that centers have more direct control of financial management from planning through execution, but it places the onus on centers to advocate funds effectively in the federal budget process.

Effective use of such funds control requires planning and coordination between users and centers. Non-price rationing of high-quality facilities can be important to implement correctly. Unfortunately, users tend to bear the brunt of any poor coordination.

When Each Center Splits the Full Annual Funding of a Facility with Its Users. Setting the transfer price such that the centers and users share funding of costs could help improve coordination between centers and users while mitigating strategic problems. This approach encourages use of multipart transfer prices that reflect marginal or activity-based costs. Prices can convey more complete information about NASA's cost structure and lead to better user decisionmaking about demands and better center decisionmaking about investments.

More complex pricing changes the administrative burden. It takes more resources to set prices and manage the fund transfers. However, it can lead to smaller differences between expected and realized revenues, since shared funding helps to cover the large, fixed overhead costs, relying on users to pay for the direct or full costs they incur for the complex for the time they use the complex.

Better information on cost structure is important to improve the objective basis for communication and shared decisionmaking. It would help centers and users work together to advocate for funds in the budget process. It also improves joint responses to surprises about cost and demand.

The Effect of FCR on the Financial Health of NASA's WT/PT Facility Portfolio

Elements of the NASA WT/PT complex shown in the bottom half of Table 4.3 (namely those at Ames) are near financial collapse. In simple terms, utilization is down because of historical and projected declines (but not eliminations) in aeronautical systems development and acquisition and because of stagnant R&D associated with the lack of a national vision, aerospace policy (national security and commercial), and supporting investments. Program utilization trends are also downward because of the current apparent dismissal of, or disregard for, the risks from reduced facility testing—even for seemingly small modifications to current robust designs. In addition, advances in CFD have reduced (but not eliminated) facility time needed for certain classes of tests. Finally, some current tunnels have limitations with respect to their ability to address certain important contemporary and anticipated aerodynamic, propulsion, and systems integration issues; upgrades to important facilities will not likely be made unless near-term financial concerns are resolved.

With declining utilization and the lack of NASA resources for (or commitment to) financial support to cover unused capacity come increased user costs. Higher user costs lead to less testing as programmatic and technical/performance risks are rebalanced. Less user testing leads to higher user costs and higher risks. If this downward utilization versus cost spiral continues, at some point it becomes unable to recover operating costs from users because of excessively high per-test costs, and the facility is lost unless other support is identified. It often takes a decade to rebuild lost capabilities. Thus, if the nation loses important facilities, there could be both serious national security and economic costs. On the commercial side, domestic industries could lose their competitive advantage to foreign firms that benefit from their governments' investment in R&D and test infrastructures. Military RDT&E costs could increase because of a reduced research base and greater reliance on high-risk and time-consuming fly-offs. Program costs will increase as schedules slip to address unforeseen problems encountered in test flights (or as programs are canceled after multibillion-dollar RDT&E costs are sunk) from a lack of insufficient earlier facility testing. In addition, opportunities to exploit revolutionary new military concepts and capabilities may be delayed while needed test infrastructures (facilities and skilled people) are redeveloped and rebuilt. Once this competitive advantage is lost, the effects on the aerospace industrial base (and hence on national security writ large) may be serious and long lasting.

Risks at Ames. The current situation at the Ames National Full-Scale Aerodynamics Complex (NFAC) facility (in the bottom-right quadrant) is a good example of the financial problem facing Ames facilities. Facilities at Ames receive no shared support and are expected to recover their full operating costs from user pricing. The NASA rotorcraft program was the major user of the NFAC but was canceled in 2002. This implied that the full cost for operating the NFAC had to be borne by the remaining user base: the Army and the few remaining sporadic testers. The Army rotorcraft program could not bear the brunt of a multimillion-dollar cost to keep the NFAC alive—partially because of short-term inflexibility in its budget and partially from the high price. Thus, the NFAC has already fallen victim to financial collapse, is currently mothballed, has seen its contractor staff let go, and is slated for full closure at the end of FY2004.

The pending loss of the NFAC is in contrast to the need for the facility. The NFAC received multiple, strong advocacies in our survey of industry, DoD, and NASA users. These users cited strategic aerodynamic needs for such a large facility despite its current utilization destabilization because of the current elimination of the NASA rotorcraft program funding.

In a recent example, the NASA Mars Exploration Rover (MER) program used the Ames NFAC WT to test the strength of the landing parachute. By using the NFAC, the Jet Propulsion Laboratory (JPL) discovered that the original design would not open fully because of "squidding."[14] JPL was able to explore and test a large

---

[14] *Mars: Dead or Alive* (2004). Squidding is a state of incomplete canopy inflation in which the canopy has a squid- or pear-like shape and the leading edges fold and curl in rather than staying fully employed (see Aircrew Survival Equipmentman 2, U.S. Navy Nonresident

number of alternative designs by using the NFAC (Ortiz, 2003) at a much lower cost and without the loss of test equipment during failed drop tests. Thus, the use of WTs for parachute testing contributed significantly to the recently successful and highly publicized MER landings. In addition to being able to observe anomalies in parachute opening that would have been catastrophic to the MER, the MER program also found that the WT tests yielded more accurate and repeatable data than traditional drop tests (Cruz et al., 2003).

Risks at Glenn and Langley. In addition, if the shared support systems in place at Glenn and Langley are disallowed by NASA's on-going interpretation of FCR policies, then those centers will also be endangered financially. Beyond FCR, there is a long-term concern that reduced aeronautic funding (and resulting reduction in the tax base used to support facilities) in the future may initiate financial crises at Glenn and Langley. It is unclear, however, from available data if and when this might occur. More than likely, it would happen first at low-utilization facilities, such as the Langley Transonic Dynamics Tunnel and Spin tunnels shown in the bottom-right quadrant.

Recovery of Full Costs Was Detrimental to Air Force Facilities. In another example, when the Air Force experimented with recovering full costs from 1969 to 1972, AEDC found that their prices became inherently unstable and unpredictable because large infra-structure-driven cost had to be spread over an annually variable customer workload base. Also, test customers were not given time to adjust their budgets to accommodate increases in testing prices. As a result, the test workload decreased dramatically (see the "Industrial Funding" era in Figure 4.2), which, in turn, drove up overhead costs and initiated a positive feedback loop that continued driving up prices and driving away users.[15] AEDC found that testing decisions

---

Training Course, NAVEDTRA #14218, 1990, available through Integrated Publishing, Spring, Texas, www.tpub.com/content/aviation/14218/css/14218_320.htm (accessed April 2004). See also White (2000) for an overview of parachutes for space-entry vehicles and problems encountered.

[15] "Feasibility of Converting AEDC to an Industrial Fund Activity," unpublished white paper by AEDC. This situation was also confirmed by the U.S. Air Force Directorate of Test

were being made based on near-term cost considerations rather than strategic considerations to reduce long-term program risks through testing. The resulting reduced testing loads and reduced income caused significant detrimental effects on AEDC's facility, including the loss of skilled people, loss of independent analysis and evaluation capabilities, decreased investments for the future, and reduced facility readiness through the loss of maintenance resources. The Air Force believes that the financial collapse at AEDC was only halted when shared support through direct budget authority was restored to AEDC.

Figure 4.2
Historical AEDC Funding

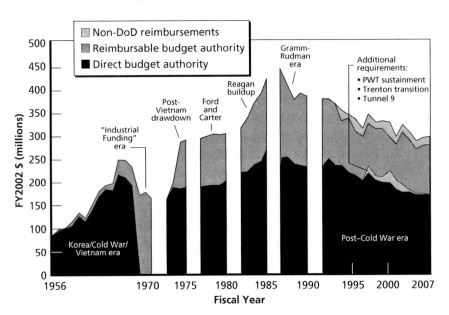

SOURCE: AEDC.
RAND MG178-4.2

and Evaluation (HQ USAF/TE), personal communication, June 2003, and AEDC, personal communication, June 2003.

Note also in Figure 4.2 that, since 1972 (when direct budget authority was reinstated at AEDC), reimbursements consistently paid for less than half of total operating costs. Thus, over many decades, the DoD has found it vital to provide shared support for its facilities despite fiscal pressures in various eras.

Summary

In looking at the condition (or health) of NASA's WT/PT facilities, two of the three key dimensions are (1) how *technically competitive* the facilities are and (2) how *well utilized* they are. By those measures, NASA's portfolio is generally in very good condition. More than three-fourths of NASA's WT/PT facilities are competitive and effective with state-of-the-art requirements. In addition, more than two-thirds are well utilized. Overall, about two-thirds are both technically competitive and well utilized, with this number varying across the categories of facilities.

However, there is room for improvement, especially in the high-Rn subsonic category and in reducing the BMAR across NASA's portfolio. There also has been discussion in the testing community for both large and small investments to improve NASA's test infrastructure, but it was difficult for our expert consultants and the user community to seriously consider large investment candidates, given declining budgets, facility closures, and the failure of past efforts to obtain funding for facilities with improved capabilities. Selected challenges, though, such as hypersonics testing, will require additional research to develop viable facility concepts for future investment consideration.

However, using a third dimension of health status—*financial health*—we find that the FCR accounting practices imposed by NASA on the centers has serious implications for the financial health of those facilities that are underutilized (about one-third of the facilities in general, with variation across the category types). Average-cost-based pricing, decentralized budgeting, poor strategic coordination between buyers and providers of NASA WT/PT facility services, and poor balancing of short- and long-term priorities inside and outside NASA are creating unnecessary financial problems that leave elements of U.S. test facility capacity underfunded. With declining usage and FCR accounting, these facilities run the risk of moving into financial collapse.

CHAPTER FIVE

How Should NASA Manage Its Wind Tunnel and Propulsion Test Facility Portfolio?

Given what we know about the alignment of NASA's portfolio of wind tunnel and propulsion test facilities (Chapter Three) and the condition (or health) of that portfolio (Chapter Four), we can next ask how NASA should manage its facility portfolio.

NASA's primary management challenges break down into two questions. First, how can NASA identify the *minimum set* of WT/PT facilities important to retain and manage to serve national needs? Those facilities that are in the minimum set should be kept, but those that are not in the minimum set could be eliminated (and, thus, constitute excess capacity from a national strategic point of view). Second, what financial concerns and resulting management steps are needed to manage the facilities within the minimum set?

Twenty-nine of 31 existing NASA WT/PT facilities constitute the minimum set of test facilities important to retain and manage to serve national needs. Thus, the test complex *within NASA* is both responsive to serving national needs and mostly "right sized" to the range of national aeronautic engineering needs. While not redundant within NASA, a few of the agency's facilities are redundant when considering the technical capabilities of the larger set of facilities maintained by commercial entities and by the federal government at DoD's AEDC. Whether these redundancies amount to the "unnecessary duplication" of facilities prohibited by the National Aeronautics and Space Act of 1958 was beyond the scope of this study. Further analysis of technical, cost, and availability issues is required to determine if facility consolidation and right-sizing across NASA and AEDC to establish a national reliance test facility plan would provide a net savings to the government. **NASA should work with the DoD to analyze the viability of a national reliance test facility plan**, since this could affect the determination of the future minimum set of facilities NASA must continue to support.

However, until the feasibility of a national reliance test facility plan is determined, **NASA should manage its portfolio to keep its minimum-set WT/PT facilities healthy and open for business. Most importantly, for those facilities in most financial danger, NASA should *identify financial shared support* to keep them from entering financial collapse because of variable utilization, FCR accounting, and lack of program support for long-term national benefits.**

In perspective, the **WT/PT facility operating costs pale in significance to NASA's overall budget and to U.S. investments in aerospace RDT&E.** NASA should keep in mind the connection between these costs and the benefits they accrue not only in the near term but to the long-term benefit of the nation's aeronautic pursuits.

The study also identified four second-order management issues and concepts that warrant mentioning for further analysis: **the importance of the test facility workforce, cross-training of facility crews, workforce outsourcing, and possible privatization options.** We discuss these findings in more detail below.

Identifying the Minimum Set of Facilities in Its Portfolio

The following four factors were used to assess which NASA facilities constitute the minimum set needed to serve national needs: alignment with national needs, technical competitiveness, redundancy, and usage.

The matrix used to summarize the health status of NASA's WT/PT facility portfolio (Table 4.3 in the previous chapter) also provides a useful tool for illustrating these three factors and their relationship in determining the minimum set. Table 5.1 uses this matrix structure and adds information on test facility alignment with national needs (including whether the facility is the primary facility serving at least one need or a facility redundant on all needs).

Determining the Minimum Set

First, facilities in the minimum set must serve national needs. Thus, facilities that no longer meet national needs are discarded from consideration out of hand.

Next, the primary NASA facilities that serve national needs are included in the set by definition. These are the primary facilities that NASA has to serve each national need. Until the need disappears or analysis can determine that it is better served outside NASA (see the discussion on collaboration and reliance below), the agency should include it in the minimum set.

Finally, facilities that are redundant to the primary facilities may or may not be included in the set depending on their technical competitiveness and utilization:

Table 5.1
Determining What WT/PT Facilities Should Be in the Minimum Set

		Weak ← Technically Competitive with State-of-the-Art Needs → Strong	
Current Usage	**Good**	Technically weak but well utilized (1 out of 31 facilities) Primary: (0/31) In the minimum set Redundant: (1/31) TBD	Technically competitive and well utilized (20 out of 31 facilities) Primary: (20/31) In the minimum set Redundant: (0/31) In the minimum set
	Poor	Technically weak and poorly utilized (4 out of 31 facilities) Primary: (3/31) In the minimum set Redundant: (1/31) Not in the minimum set	Technically competitive but poorly utilized (6 out of 31 facilities) Primary: (6/31) In the minimum set Redundant: (0/31) TBD

- Technically competitive and well-utilized redundant facilities (upper-right quadrant) provide important capacity supplements to primary facilities and should therefore be included in the minimum set. No current NASA WT/PT facility fell into this category.
- Technically weak and poorly utilized redundant facilities (lower-left quadrant) are truly excess capacity from a strategic perspective and should not be included in the minimum set. Only one current NASA WT/PT facility fell into this category: the Langley 12-Foot Atmospheric Laboratory.
- The borderline cases come with the redundant facilities in the upper-left (well utilized but technically weak) and lower-right (technically competitive but poorly utilized) quadrants. The status of these facilities is marked as "TBD" (to be determined), since additional insights into their relative technical competitiveness, the potential for long-term need for those facilities, and their competitors are needed to resolve whether or not they should be included in the minimum set on a case-by-case basis. Only one current NASA WT/PT facility fell into either of these categories: the Langley16-Foot Transonic Tunnel (16TT).

As a result of this analysis, only two existing NASA WT/PT facilities are candidates for exclusion from the minimum set.

The Langley 12-Foot Subsonic Atmospheric Laboratory is a technically weak and poorly utilized redundant facility in the lower-left quadrant. It has very-low-Rn capabilities but has been inexpensive to operate and readily available for NASA research. The Langley 14×22-Foot is a larger, inexpensive alternative (although Langley management indicated that testing costs in its 12-Foot are about 1/15 that of the still inexpensive 14×22-Foot). The Langley 12-Foot has poor technical competitiveness and poor utilization. More specifically, the 12-Foot has poor user advocacy outside Langley and is viewed as unacceptable by important industry segments. It has a poor utilization history and poor forecasted future need. Therefore, the Langley 12-Foot laboratory should be excluded from the strategic minimum set.

The Langley 16TT is well utilized but technically weak. The 16TT has been affordable, available, and highly used, but its poor technical capabilities make it less competitive against state-of-the-art requirements. However, the 16TT has low-Rn capability, poor flow quality, and a low maximum Mach number. Thus, it is inferior to the Ames 11-Foot as well as the DoD's AEDC 16-Foot Transonic Wind Tunnel (even though the 16TT costs less to operate than the larger pressure tunnels do[1]). Therefore, the Langley 16TT should also be excluded from the strategic minimum set.

It is important to note that the most important basis for inclusion in the minimum set was not utilization but whether a facility had a primary role in serving a strategically important national need. Utilization is a useful factor in determining the inclusion or exclusion of redundant facilities, but when considering the primary need server (or the only remaining facility serving a national need), utilization should take a back seat to strategic need.

Managing the Facilities in NASA's Portfolio

The actions NASA should take to manage a WT/PT facility depend on where that facility falls within the matrix in Table 5.1. Table 5.2 shows how NASA should manage the WT/PT facilities within the four quadrants of the matrix.

Managing WT/PT Facilities Outside the Minimum Set

First, let us consider the management options for facilities outside the MS. In general, those facilities lack a strategic need but may have other near-term needs or needs that fall outside those of national importance. Thus, non-MS facilities could be left to survive on their own without any shared support, or they could be mothballed or

---

[1] It should be noted that operating cost discussions do not consider test pricing approaches or the effects of subsidies or shared support on pricing. They are general reflections of operating costs for the facility, regardless of how the operating costs are recovered.

Table 5.2
A Way for NASA to Manage Its WT/PT Facility Portfolio

		Weak	Strong

Current Usage	**Good**	Technically weak but well utilized (1 out of 31 facilities)
		
		Primary: (0/31)
		  Minimum set: upgrade or replace
		Redundant: (1/31)
		  Minimum set: upgrade or replace
		  Not: Close or on own
			Technically competitive and well utilized (20 out of 31 facilities)
			
			Primary: (20/31)
			  Minimum set: Provide *some* shared support
			Redundant: (0/31)
			  Minimum set: Provide *some* shared support

	Poor	Technically weak and poorly utilized (4 out of 31 facilities)
		
		Primary: (3/31)
		  Minimum set: Upgrade or replace
		Redundant: (1/31)
		  Not: Close or on own
			Technically competitive but poorly utilized (6 out of 31 facilities)
			
			Primary: (6/31)
			  Minimum set: Provide *significant* shared support
			Redundant: (0/31)
			  Minimum set: Provide shared support
			  Not: Close or on own

Technically Competitive with State-of-the-Art Needs

sold. Lacking sufficient need and given ample savings to be gained, the facility could be closed.

The two NASA facilities not included in the minimum set were the Langley 16TT—in the upper-left quadrant—and the Langley 12-Foot Subsonic Laboratory—in the bottom-left quadrant. Thus, both are assessed as technically weak but differ in their utilization.

The Langley 16TT is already slated for closure by NASA. It might be prudent for NASA to mothball the 16TT for the long term until it is obvious that no requirements will develop in subsequent strategic plans and then abandon (close) the facility.

The Langley 12-Foot Subsonic Atmospheric Laboratory has very-low-Rn capabilities, but it has been very inexpensive to operate and readily available for NASA research. The Langley 14×22-Foot is a larger, inexpensive alternative (although Langley management indicated that testing costs in the Langley 12-Foot are about 1/15 that of the still inexpensive 14×22-Foot). The Langley 12-Foot has poor technical competitiveness and poor utilization. More specifically, the Langley 12-Foot has poor user advocacy outside Langley and is viewed as unacceptable by important industry segments. It has a poor utilization history and poor forecasted future need. All told, the Langley 12-Foot has a poor strategic position, yet very low operating costs. As such, it would be candidate for closure.

Managing WT/PT Facilities Inside the Minimum Set

Management options for WT/PT facilities within the minimum set vary depending on the technical competitiveness and utilization of each WT. Since the matrix quadrants are based on these two factors, the quadrants help us to understand what management actions are applicable in dealing with the facilities that fall in these quadrants.

Currently, the remaining 29 NASA WT/PT facilities are all assessed to be within the minimum set. Those facilities—or the capabilities they provide—need to be maintained as follows.

Technically Competitive and Well Utilized. The WT/PT facilities that fall into the upper-right quadrant—technically competitive and well utilized—are in fairly good health at present and must be managed to ensure that they stay that way. The key here is providing shared support as needed to keep these facilities from falling down into the bottom-right quadrant if utilization begins to diminish and, ultimately, toward financial collapse.

Currently, 20 NASA WT/PT facilities fall into this category.

Technically Competitive but Poorly Utilized. As with the upper-right quadrant, the WT/PT facilities in the minimum set that fall into the lower-right quadrant—technically competitive but poorly utilized—are in fairly good health at present and must be managed to ensure that they stay that way. Unfortunately, their poor utilization introduces a financial challenge to management and should be addressed with shared support (as discussed in Chapter Four) to keep the facilities from falling behind technically or from financial collapse.

Since these facilities are poorly utilized, an additional management concern is the maintenance not only of the test facility hardware but of the workforce. If the facility is not periodically used, then workforce skills will degrade and have a negative effect on its technical competitiveness.

Currently, six NASA WT/PT facilities fall into this category.

Technically Weak Regardless of Utilization. Technically weak facilities that lie in the minimum set require investments in the form of upgrades or, in some cases, replacement to make NASA's capability technically competitive.

A lesson learned from the recent rebuild of the Ames 12-Foot Pressure Wind Tunnel was the possible loss of the user base during extended closure for upgrades. This concern is especially relevant for facilities that are currently in high use.

With the poorly utilized facilities, an additional management concern is the upgrade of not only the test facility hardware but of the workforce. Continued low utilization without investments to maintain the workforce could further degrade the technical competitiveness of these facilities.

Currently, no NASA WT/PT facilities in the minimum set are technically weak but highly utilized, and three are technically weak and poorly utilized. (The other two facilities in the technically weak column are not in the set.)

Options for Managing WT/PT Facilities in the Minimum Set

Here, we examine some options for managing WT/PT facilities in the minimum set, particularly for those that are underutilized. These options (discussed below) include financial shared support, collaboration and reliance options, and ownership transferal options.

Financial Shared Support. Despite their importance, NASA facilities requiring financial shared support are unhealthy and require the immediate attention of NASA leadership. Despite some technical competitive issues discussed earlier, the most pressing health concern facing NASA facilities is the unreliable and dwindling funding stream to keep these facilities open and well maintained, especially in periods of low utilization.

As noted in the previous chapter, NASA's WT/PT facilities are currently managed in a decentralized approach in which the three autonomous NASA research centers that own and operate them

(Ames, Glenn, and Langley) set their own directions and management processes for the facilities located in their centers. NASA has required that the centers adopt FCR accounting methods. These management problems are more than just accounting issues. Accounting for operating expenses is one issue, but how those costs are recovered (e.g., by NASA subsidies or passed on to users) is the real issue. Generally, low current utilization results in a low-income stream. If NASA does not identify significant shared support for these facilities, it risks sending them into financial collapse, as discussed in the previous chapter.

Given the three basic approaches for sharing funding and the surrounding issues outlined in the previous chapter, NASA should use shared funding of annual full costs to set transfer prices. This approach allows centers and users to split budgetary burden. It makes the need for long-term coordination between users and centers more apparent, ensuring the long-term health of strategic facilities. It allows centers and users to tailor coordination by facility. It fairly asks the users to pay the costs of the testing they conduct while ensuring that the fixed overhead of maintaining strategically important facilities is not at the mercy of year-to-year variances in need or closed when the user base falls below some critical mass in a given year (or even multiyear trend period). Transfer prices can be tailored to support better user and center decisionmaking, and contingency funds can be established to cover the difference between expected and realized revenues at the end of the year.

Together with the need to better account for the full costs of test facilities, the DoD came to similar conclusions when it established the major range and test facilities base (MRTFB) and advocated that users need to see the cost they impose on a facility while not being asked to pay for unused and underutilized capacity at the strategically important test facilities they use.[2]

NASA has indicated that FCR does not imply that all costs must be recovered through direct test pricing. Thus, even the implementa-

---

[2] See Bergquist et al. (1972); DoD Directive 3200.11 (2002); and DoD 7000.14-R (1998).

tion of FCR by NASA headquarters on test facilities can include the provision for shared support in setting transfer prices.

Furthermore, NASA policy states that "[w]here a NASA activity provides special benefits to an identifiable recipient above and beyond those which accrue to the public at large, a charge shall be imposed to recover at least the full cost to the Federal Government of providing that service, goods, or resource."[3] Providing shared support to keep strategically important facilities alive and healthy when not being utilized provides "special benefits" that "accrue to the public at large" and therefore need not be charged to users.

Financial Sources for Shared Support. A challenging task facing NASA leadership is identifying resources to provide shared support for its WT/PT facilities. Facilities strategically important to an aerospace sector require such funding during low-utilization periods so that users continue to pay only for stable, reasonable costs.

Shared Support from NASA. The most straightforward sources for shared support are from the local research center, from NASA headquarters, or from a program. Langley and Glenn currently provide shared support, as described in the previous chapter, through overhead mechanisms on local research programs. This support is in danger of FCR policy decisions, but, again, as discussed in the previous chapter, those policy decisions should be crafted to avoid endangering shared support.

Another approach for shared support would be for NASA headquarters to create a budget line for such purposes and administer the support centrally. This would avoid the problem exhibited at Ames, where a local research center's focus has reduced the emphasis on test facilities, resulting in a lack of shared support from the center. NASA headquarters has indicated that OMB is against establishing direct budgets for test facilities, but we feel that this position should be reconsidered given that centrally funded support (as provided by the DoD to the MRTFB) may be needed in certain circumstances.

---

[3] See NASA (1999, paragraph 1a).

A third approach would be the establishment of a research program for test technologies to include shared support for test facilities. This approach is perhaps the most desirable within NASA, since such a program could not only provide a lifeline for facilities but also address a need widely expressed by users that NASA continue to advance the state of the art in test technology. Such investments would therefore provide funding for data acquisition technology, advancement in CFD, exploration of testing needs, research into testing techniques such as hypersonic propulsion integration testing, upgrades to existing facilities, and the development of well-documented and grounded advocacy for new facilities. Such a program should be administered in a way so that shared support is not restricted to a single research center but can be flexibly applied to whatever facility is in need, regardless of location.

Shared Support from a Consortium or Retainer. Shared support does not have to come solely from NASA. The DoD and industry are consolidating or closing facilities to save money because of lower utilization; thus, NASA could solicit their support for NASA-owned facilities as a "win-win" way to use some of those savings to ensure the health of the remaining facilities. NASA should engage the DoD and industry in discussions to investigate the possibilities.

Similar to a consortium, NASA could engage users one-on-one to explore the possibility of a multiyear retainer for its facilities. Boeing, for example, is ensuring its access to QinetiQ 5M through retainer agreements.

Collaboration and Reliance Options. While most NASA WT/PT facilities are used for NASA research, a few other of the agency's facilities are currently (but not exclusively) dominated by RDT&E related to military vehicles. While these facilities have historically been used by a mix of military and civil testing, consideration could be given to putting them under the control of the DoD, unless NASA identifies sufficient general aeronautical research or industry support to retain these facilities in the broader civil domain. Specifically, some of these NASA facilities provide capabilities that could be met by some (often larger and more expensive to build and operate) test facilities maintained by DoD at AEDC or by foreign facilities. Whether it is prudent to close selected NASA facilities and rely on these other facilities requires further analysis.

It is unknown, for example, whether further facility consolidation across NASA and AEDC would provide a net savings to the government. Multifacility sites have large, common infrastructures with fixed recurring costs that are not reduced when a subset of the site's facilities are closed; closures merely shift these fixed infrastructure costs to the remaining facilities. Also, gross savings from the elimination of a facility's operating budget must be weighed against increased costs in time, travel, shipping, higher test costs, and lost opportunity to research programs to test at nonlocal facilities.[4] There is also the risk that unforeseen future programs may need facilities whose unique capabilities appear less important today. Cost data to understand the financial implications of these testing trade-offs were not available during this study, primarily because NASA was still implementing full-cost accounting of its operations. Thus, the need to understand these costs is a major reason why NASA should continue to pursue full-cost accounting to inform current and future facility management and reliance decisions. Other data such as alternative facility testing quotes and costs to programs for nonlocal testing would require additional, in-depth analysis across the facility alternatives and NASA programs.

Beyond cost considerations, reliance on facilities outside NASA requires a clarification of availability as well as an examination of needed resident expertise. Currently, DoD programs have first priority at DoD facilities, potentially restricting NASA or industry access for extended periods. NASA and the DoD would need to discuss such access issues and determine whether research programs could forgo access for such periods or whether surge capabilities are an

---

[4] It is unclear how strong a requirement co-location of a tunnel with a research community is. Convenience is certainly a factor in advocacy for local facilities, and transportation costs can be significant. However, distant facilities may offer better capabilities, and some advances in remote data monitoring have reduced some needs to send the full test crew to distant sites. Industry users are accustomed to testing at nonlocal sites because of necessity. A full analysis would be needed for each candidate set of tunnels to understand how the costs compare with the benefits.

option (e.g., by adding extra shifts). Also, tests at facilities located at NASA's Glenn and Langley benefit from the resident aeronautic expertise at these centers, so an examination should be made to understand how the lack of such on-site expertise at AEDC would affect the quality of R&D performed at AEDC facilities.

Candidates for collaboration and reliance on DoD facilities include those listed in Table 5.3.

The consolidation and reliance issues are not straightforward. In many cases, the AEDC facilities provide more capabilities than are needed, and they are more expensive to operate for a given test. In other cases, there are technical differences between the facilities that may preclude trade-offs (e.g., open-loop propulsion exhausting at Glenn propulsion wind tunnels compared with closed-loop scoop exhaust recovery at AEDC facilities[5]).

Table 5.3
Technically Similar Facilities in NASA and AEDC

NASA Facilities	Technically Similar AEDC Facilities
Ames 11-Foot Transonic Wind Tunnel	AEDC 16-Foot Transonic Wind Tunnel
Glenn 8×6-Foot Propulsion Transonic Tunnel	AEDC 16-Foot Transonic Wind Tunnel
Glenn 10×10-Foot Propulsion Supersonic Tunnel	AEDC 16-Foot Supersonic Wind Tunnel
Glenn Propulsion Simulation Laboratory	AEDC C, J, and T cells
Ames 12-Foot Pressure Wind Tunnel	It may be possible to modify the AEDC 16-Foot Supersonic Wind Tunnel to provide efficient high-Rn subsonic testing at a reasonable cost. Further technical and cost analysis is required.

---

[5] For example, for models that introduce contaminants into the air stream (e.g., engine or model exhaust tests) in the Glenn 10×10-Foot Supersonic Tunnel, the facility is configured to exhaust all air (in an open-loop configuration using a valve), rather than returning the air over the test section, to ensure that no exhaust components return to the engine or test model. At AEDC, an adjustable scoop system is used to extract contaminants from the air stream while returning the remaining air to the model in a closed loop. In many current cases, scoop appears to be satisfactory, but analysis of all test configurations is needed to understand what future tests may be affected by the scooping approach.

Foreign reliance opportunities are also available. The most significant option is in the high-Rn, general-purpose subsonic WT category. The Ames 12-Foot Pressure Wind Tunnel is the only U.S. capability in this category and has been historically important for civil, space, and military vehicle RDT&E. Currently, however, the Ames 12-Foot Pressure Wind Tunnel has poor competitiveness against needs by important industry segments and has very poor utilization. Thus, the Ames 12-Foot Pressure Wind Tunnel could be removed from the NASA minimum set *if* the costs associated with NASA testing in Europe are satisfactory and availability arrangements could be made. Anecdotal indications are that QinetiQ is looking for additional reliable usage for its facility and would probably welcome reliance discussions with NASA. More important, however, such a reliance arrangement *assumes that the United States wants to rely on a foreign capability for such a strategic test capability* important not only for the commercial sector but for military vehicles important to national security.

Ownership Transferal Options. In addition to reliance options, certain NASA facilities are currently dominated by military RDT&E and may be candidates for transferal to DoD support or ownership. NASA facilities in this category include the Ames NFAC, Langley Spin Tunnel, and Langley Transonic Dynamics Tunnel. However, such transactions may not save the government money, since support would come from DoD rather than NASA. It is also unclear whether these facilities are so extensively operated for military purposes to warrant such actions. The NFAC is an important facility for civil rotorcraft testing, and until the recent demise of the separate NASA rotorcraft program, NASA used it extensively. Indications are that other NASA research programs may be picking up some rotorcraft research, and Congress has considered reinstating the rotorcraft program.[6] The Langley Spin Tunnel and Transonic Dynamics Tunnel have been used on a number of commercial aircraft programs, and

---

[6] See, for example, Sec. 102, "Rotorcraft Research and Development Initiative," of U.S. Senate (2003).

the aeronautics field uses the lessons learned by military RDT&E in those facilities.[7]

Cost Perspectives

In thinking about issues of shared financial support for NASA's WT/PT facilities, it is important to retain perspective on the magnitude of NASA's facility costs relative to the investment value of the vehicles they enable or support. In FY2002, the annual operating budgets for all NASA test facilities under study total approximately $125–130 million (depending on the accounting standards used).[8] Non-NASA users reimburse approximately $20 million of this operating budget. Thus, these test facilities comprise approximately $100 million a year—less than 5 percent of the 2002 Aerospace Technology budget of about $2.5 billion, and less than 1 percent of the total NASA budget of about $15 billion, in the 2002 Initial Operating Plan.[9]

These operating budgets are relatively small compared with the total investments made in producing aeronautic vehicles. For example, federal investments in aerospace R&D have totaled between $32 billion and $58 billion annually in the past decade but with a downward trend. Military aircraft RDT&E funding alone had run from $4.5 billion to $7 billion in the same period, showing anticipated upswings in the near future.

---

[7] Chambers (2003).

[8] The FY2002 operating budget figure of $125–130 million was derived from an amalgamation of the information briefed to RAND during the team's site visits and the interim results of the National Aeronautics Test Alliance (NATA) Cost Panel as of November 7, 2002. This uncertainty of knowing what it costs to operate NASA's facilities is a strong reason supporting NASA's drive for full-cost accounting to be implemented in FY2004. Even if different centers adopt different accounting standards, it is important to be able to routinely know what the actual costs are and what standards were used to obtain those figures.

[9] By FY2004, NASA reorganized and separated the space research from the rest of the aeronautics research. In the *FY 2004 Initial Operating Plan* (dated March 10, 2004), the Aeronautics Research Budget amounted to $1.037 billion out of a total NASA budget of $15.378 billion.

Insights into Alternative Management Models

While financial shared support for important low-utilization facilities was deemed the most pressing management problem facing NASA, the study did identify a few second-order management issues and concepts that warrant mentioning for further analysis and consideration.

Importance of the Test Facility Workforce

A significant theme from users and test facility operators alike was the importance of a knowledgeable, skilled, and motivated workforce to complement the technical hardware of the facility. *Hardware alone is not enough*. Users also noted the value of having active researchers on hand at NASA R&D facilities as a major advantage. The Glenn Icing Research Tunnel is a prime example of the strong advantage of having seasoned and knowledgeable operating crews along with researchers on hand to address testing issues.

Cross-Training of Crews

Training crews so they can operate more than one facility has reduced operating costs in NASA and the DoD.[10] Heavy cross-training of crews at Ames has enabled the interesting concept of managing the test complex as a set of test sections rather than separate WT/PT facilities; they coined the phrase "two crews, five test sections" to describe their operating paradigm (although some dedicated staff members remained full-time to keep certain facilities such as the NFAC operational and to retain their skills). This flexibility allowed the test complex management to somewhat balance the variance of facility utilization across the whole center (until the lower-than-expected utilization and effects of full-cost recovery with no shared support forced the recent decisions to close the Ames NFAC and 12-Foot tunnels).

---

[10] Personal communications with Langley, Glenn, Ames, AEDC, and Air Force Research Laboratory (AFRL) wind tunnel facility managers, 2002.

Other organizations have also employed cross-training to better utilize staff time. AFRL said that it also uses cross-training extensively to maintain skills in the small crew it has across the laboratory.

Langley has implemented some cross-training and crew-shifting, but nonuniform computer equipment and some concerns from at least one accident resulting from sharing inexperienced staff have hindered Langley's efforts. Information technology investments to standardize the center's data processing equipment would facilitate greater implementation. Determining the actual returns on such investment requires further analysis.

Workforce Outsourcing

Finally, NASA (as well as DoD at AEDC) is increasingly outsourcing its test facility workforce (i.e., transitioning the workforce from civil service to contractors) to obtain greater flexibility in hiring, relocating, and, in some cases, reducing excess staff. In many cases, the same experienced, former civil-servant individuals were retained under contract services. AEDC management has been very happy with its contract workforce and 90/10 split between contract support and civil-servant management. Although additional analysis is required to fully understand the implications, crew outsourcing appears beneficial with one critical caveat: The civil-service cadre that provides the leadership and corporate memory for the test facility enterprise must be sustained at a critical mass to ensure that the government can provide sound direction and investment decisions as a wise buyer of contractor services.

A Privatized Operations Model

In the long term, NASA might explore alternative management constructs for its facilities. One such construct is the facility operations outsourcing model recently enacted by the UK Ministry of Defence (MOD) for its T&E facilities.

Under this model, the MOD identified the T&E facilities it needed for the future[11] and then privatized the operation of those facilities. Ownership of the fixed equipment and land were retained by the MOD for indemnity reasons, but ownership of the movable items was transferred to a private company, QinetiQ.[12]

Under the model, QinetiQ has a 25-year contract for operating the facilities. The contract is structured to encourage the company to implement efficiencies while retaining the long-term health and availability of the facilities. QinetiQ gets to keep the efficiency savings realized during the then-current five-year period of the contract. When the next five-year period is negotiated, the MOD receives the benefits of the efficiencies by adjusting the period funding amount to the new efficiency level.

The key to this model was the MOD's access to all facility costs to ensure that support levels in the contract guarantee the success of QinetiQ *and* the facilities. In U.S. parlance, this would require open full-cost accounting not only of the facilities but also of the acquisition programs that rely on the facilities.

The model employs shared support for the facilities. In the current five-year period, the MOD centrally funds 84 percent of facility costs to keep the facilities' doors open. Sixteen percent of the funding comes from programs to support direct costs of specific program test activities. In the past, the MOD had to query programs for how much they would put in first. Now, the health of the facilities are ensured and planned for, regardless of the realized utilization in the known-variable environment.

---

[11] As it turns out, the United Kingdom decided that WTs were no longer strategically important, since the country made the conscious decision to stop building military airframes. Nevertheless, we feel that this model provides interesting insights on how T&E facilities that remain important can be managed in an efficient and cost-effective way while ensuring their health.

[12] QinetiQ is a public-private partnership between the MOD and the Carlyle Group. QinetiQ is a British company based in the United Kingdom. The MOD retains a "special share" in the business to ensure that the United Kingdom's defense and security interests continue to be protected. Safeguards were installed to prevent conflicts of interest and to ensure that the integrity of the government's procurement process is not compromised. See QinetiQ's Web site at www.qinetiq.com for more information.

The MOD implemented the model in 2003, so the success of it has yet to be established. Nevertheless, some important observations can be made. The MOD made a conscious, objective decision about which facilities are strategically important in the long term (25 years, in this case). The ministry ensured that it accounted for all the costs to inform its decision. It provided shared support for the facilities to ensure their long-term health, independent of the yearly utilization. While not having to perform the actual operation of the facilities, the MOD provided controls and incentives to realize efficiency and cost savings while ensuring quality and availability of needed facilities. NASA could learn from these observations.

Summary

While there is certainly significant unused capacity in most NASA facilities, *testing cannot be consolidated in a significantly smaller set of NASA facilities to reduce costs.* As such, 29 of the 31 WT/PT facilities represent a minimum set of capabilities, which means they are unique (within NASA) in serving at least one national need. Thus, the current type of test complex *within NASA* is both responsive to user needs and is mostly "right sized" to the range of national aeronautic engineering needs.

While utilization was used in the previous chapter to help inform competitiveness, it is important to recognize that, given situations of unique facilities meeting important national needs, utilization is best viewed as only one measure of the current financial income for a facility. When alternate facilities are not available and program or production rates continue but at a slower rate, utilization is *not* a good measure of excess capacity.

The key management challenge for NASA is to identify shared support to keep its minimum set of facilities from entering financial collapse due to variable utilization, FCR accounting, and lack of program support for long-term national benefits. It is important to note that the approximately $125–130 million annual operating budgets for all NASA WT/PT facilities under study pale in significance to the national aerospace capabilities that they partially enable, including the federal investments in aerospace R&D that have totaled between $32 billion and $58 billion annually in the past decade (with a downward trend)[13] and the military aircraft RDT&E funding alone that has run from $4.5 billion to $7 billion per year in the same period.

Within NASA, the primary facility management problem relates to funding the three autonomous research centers' test facilities in the face of declining R&D budgets. In the extreme case at Ames, the lack of resident aeronautics research programs, combined with the center management's strategic focus toward information technology and away from ground test facilities, has left Ames WT/PT facilities without support beyond user testing fees. Thus, the Ames facilities are vulnerable to budgetary shortfalls when utilization falls. Two Ames facilities that are unique and needed in the United States have already been mothballed as a result. The other NASA research centers with WT/PT facilities—Glenn and Langley—rely heavily on resident research program taxes to cover low-utilization periods in their major test facilities, but NASA center managers do not yet know whether full-cost recovery policies will nullify these funding sources.

If NASA management is not proactive in providing financial support for such facilities beyond what is likely to be available from FCR pricing, then the facilities are in danger of collapsing financially. In the near term, this market-driven result may allow NASA to reallocate its resources to serve more pressing near-term needs at the expense of long-term needs for WT/PT facilities. Given the continuing need for the capabilities offered by these facilities for the RDT&E of aeronautic and space vehicles related to the general welfare and security of the United States, the right-sizing NASA has accomplished to date, the indeterminate costs to decommission or

---

[13] Federal aerospace procurements and R&D expenditures in the period of FY1993–FY2001 ranged from a high of $58 billion in FY94 to a low of $32 billion in 2001 (Source: RAND research by Donna Fossum, Dana Johnson, Lawrence Painter, and Emile Ettedgui, published in the *Commission on the Future of the United States Aerospace Industry: Final Report* [Walker et al., 2002, pp. 5–10]).

eliminate these facilities, the significant time and money that would be required to develop new replacement facilities, and the relatively modest resources required to sustain these facilities, care should be taken to balance near-term benefits against long-term risks. Collaboration, reliance, and ownership transferal options for obtaining alternative capabilities in lieu of certain facilities are possible, but even if these options are exercised, many facilities will remain unique and critical to serving national needs.

The management solutions to address long-term priorities hinge in most part on the dedication of financial resources to preserve important facilities through multiyear periods of low utilization. Management options in terms of who owns and who operates the facilities (e.g., government or private; NASA, DoD, or confederation; NASA-center-centric or centralized) will have various pros and cons, but all will require a mechanism to stabilize and preserve capabilities needed in the long term through lean times. Key to subsequent analysis of these options is the collection and availability of the full costs of operating these facilities as well as the full costs associated with relying on alternative facilities. This report will help provide the motivation to address these policy, management, and cultural problems, ensuring the continued health of the nations' civil, military, and commercial aeronautics enterprises.

The study also identified a few second-order management issues and concepts that warrant mentioning for further analysis consideration: the importance of the test facility workforce, cross-training of facility crews, workforce outsourcing, and possible privatization options.

CHAPTER SIX

Options and Recommendations

NASA is currently capable of providing effective quality support to its WT/PT facility users within and outside NASA in the near term. However, the agency must now begin to address more proactively some important technical and management issues and potentially adverse trends to stabilize the current situation and address long-term state-of-the-art testing requirements. If it does not act, serious deficiencies may emerge in the nation's aeronautics R&D and T&E capabilities over the next 10 to 20 years. Proactive approaches to mitigate these potential problems have both management and technical dimensions.

In this final chapter, we discuss essential elements of our approach, followed by our recommendations.

How the Approach Frames the Conclusions and Recommendations

The analytic method used in the study to define needs does not rely on an explicit national strategy document for aeronautics in general and for WT/PT facilities in particular because it does not exist. Lacking such an explicit needs document, we examined what categories of aeronautic vehicles this country is currently pursuing, plans to pursue, and will likely pursue based on strategic objectives and current vehicles in use.[1]

Also, as *enabling infrastructures*, WT/PT facility operations are not funded directly by specific line items in the NASA budget.[2] The study's determination of facility needs and the resulting conclusions and recommendations are, therefore, not based on the federal budget process as a direct indicator of policy dictates of WT/PT facility need. Facility need was determined by identifying what testing capabilities and facilities are required given current engineering needs, alternative approaches, and engineering cost and benefit trade-offs. This, of course, can lead to a bias in the findings because these assessments may overly reflect what the engineering field determines is important rather than what specific program managers are willing to spend on testing due to program budget constraints. Thus, when a needed facility is closed because of a lack of funding, there is a disconnect between current funding and prudent engineering need, indicating that the commercial and federal budget processes may be out of step with the full cost associated with research and design of a particular vehicle class, and a lack of addressing long-term costs and benefits.

Policy Issues, Options, and Recommendations

Table 6.1 lays out and summarizes the main policy issues identified in the study, along with the decision space for those issues and our assessment of the viability of the options. Nearly all options are either

---

[1] Specific projects and plans were obtained from NASA, Office of Aerospace Technology (2001; 2002); NASA (2001; 2003); The National Aeronautics and Space Act of 1958; DoD (2000; 2002); FAA (2002); NRC (2001); Walker et al. (2002); NASA, Office of the Chief Financial Officer (n.d.); AFOSR (2002); and various DoD and commercial research and production plans.

[2] The *construction* of government WT/PT facilities are, however, very large expenditures requiring explicit congressional funding, and certain facilities such as the National and Unitary facilities have associated congressional directives regarding operation and intent.

Table 6.1
Key Policy Issues, Options, and Recommendations

Strategy	Technical Need	Facilities Needed		Operating Costs	Investments
Issue: How should testing be addressed strategically?	Issue: How should aeronautic testing needs be met?	Issue: Which facilities should NASA keep in its minimum set?	Issue: What to do with low-utilization facilities?	Issue: How should the facilities be funded?	Issue: What future investments should be made?
1. Develop a NASA-wide aeronautics test technology vision and plan **2. Work with the DoD to analyze the viability of a national test facility plan**	1. Replace facility testing with CFD 2. Replace facility testing with flight testing 3. Use facility testing exclusively **4. Use appropriate mix of facility, CFD, and flight testing**	1. Keep all facilities that align with national needs **2. Leave out aligned facilities that are weakly competitive, redundant, and poorly utilized** 3. Leave out all facilities technically similar to those in the DoD	1. Close when utilization is low 2. Mothball when utilization is low **3. Reassess long-term needs and keep those strategically important** 4. Keep all facilities regardless of utilization	1. Recover all costs from users **2. Share support between users and NASA institutional funding** 3. Pay all costs through institutional funding so testing is free to users	**1. Eliminate backlog of maintenance and repair (BMAR)** **2. Conduct hypersonics test facility research** 3. *Pursue high-productivity, high-Rn subsonic and transonic facilities* **4. Continued investments in CFD research**

NOTES: **Bold = Recommended**; *Italics = Unclear*; Roman = Not Recommended.

specifically recommended or not recommended. One non-recommended option related to investments could be pursued, but it is unclear how viable it is in today's financial climate.

Note that the issues and options tend to be interrelated. For example, the determination of which facilities are important to keep is related to the question of what to do with low-utilization facilities. The recommended options are also related. For example, developing a long-term vision and plan for aeronautic testing, reviewing the technical competitiveness of facilities, and sharing financial support for facilities with users are interrelated.

An Overarching Issue: How Should Testing Be Addressed Strategically?

Our study identified and recommends two policy options to strategically address aeronautic testing needs for the United States.

Develop a NASA-Wide Aeronautics Test Technology Vision and Plan

The most critical issue is for NASA headquarters leadership to **develop a specific and clearly understood aeronautics test technology vision and plan**; this strategy should address the underlying issues by continuing to support the development of plans to very selectively consolidate and broadly **modernize existing test facilities**, and proscribing **common management and accounting directions** for NASA's WT/PT facilities. This vision cannot be developed in isolation from other critical decisions facing the nation. It must be informed by the aeronautic needs, visions, and capabilities of both the commercial and military sectors supported by NASA's aeronautical RDT&E complexes. Currently, the national aeronautical strategy appears disconnected from the funding that OMB and Congress provide to the users and operators of WT/PT facilities.

The NASA aeronautics vision as outlined in its Aeronautics Blueprint presents some interesting concepts, but it is not clear how they would affect the need for, and capabilities of, future test infrastructures. And the development of test capabilities per se is not an explicit part of the NASA vision. Of course, a vision is hollow without supporting investments, and aeronautics is a decreasingly significant part of NASA relative to space investments (which may become even smaller in response to budget pressures).

Given the inherent inability to reliably and quantitatively predict all needs for RDT&E to support existing programs much beyond a few months out and the trends indicating a continuing decline in needed capacity to support these needs for the foreseeable future, **long-term strategic considerations must dominate.** If this view is accepted, NASA must find a way to sustain indefinitely, and in a few cases, to enhance, its important facilities (or seek to ensure reliable and cost-effective alternatives to NASA facilities) as identified in this study.

Work with the Department of Defense to Analyze the Viability of a National Test Facility Plan

While mostly not redundant within NASA, a few of NASA's facilities are redundant when considering the technical capabilities of the larger set of facilities maintained by commercial entities and by the federal government at DoD's AEDC. Whether these redundancies amount to the "unnecessary duplication" of facilities prohibited by the National Aeronautics and Space Act of 1958 was beyond the scope of this study. Further analysis of technical, cost, and availability issues are required to determine whether WT/PT facility consolidation and "right-sizing" across NASA and AEDC would provide a net savings to the government. **NASA should work with the DoD to analyze the viability of a national test facility plan,** since this could affect the determination of the future minimum set of facilities NASA should continue to support.

While this study addressed the question of needed test facilities from a *NASA-centric* facility perspective, these results (while incomplete from a national perspective) will provide a useful starting point in future studies that may attempt to rationalize all government WT/PT facility capabilities. However, since many DoD facilities are larger and more expensive to operate for a given test than are comparable NASA facilities, and since NASA's focus is on commercial and civil aeronautics R&D whereas DoD's focus is on military systems RDT&E, national-level federation may not be desirable. In any event, such an analysis will require, among other things, data on the full cost of operating each facility; this information was not available during our study because NASA was still working to implement full-cost accounting with a target of FY2004. Such analysis will also require detailed assessments of additional costs to government programs, including firm price quotes from alternative test facilities, detailed assessments of travel costs incurred from nonlocal testing, and analysis of the stability and availability of alternative facilities.

How Should Aeronautic Testing Needs Be Met?

The nation continues to need general-purpose R&D and T&E testing facilities across all speed regimes, as well as for specialty tests. These facilities advance aerospace research, facilitate vehicle design and development, and reduce design and performance risks in aeronautic vehicles. But how should these testing needs be met?

While CFD has made inroads in reducing *some* empirical test requirements capabilities, the technology is not yet reliable for predicting the characteristics of the complex separated flows that dominate many critical design points for an aircraft. Thus, CFD will not replace the need for test facilities for the foreseeable future.

Flight testing plays a dominant role during final refinement, validation, and safety verification of a production aircraft, but except for small vehicles for which multiple, full-scale, flight-capable vehicle concepts can be quickly and relatively inexpensively produced, flight testing complements but does not replace facility testing.

Test facilities alone cannot answer all testing questions. CFD has proven an excellent and cost-effective tool for some situations such as preliminary design configuration screening. Some flight conditions and vehicles cannot be economically simulated in WT/PT facilities.

Thus, all three approaches are *complementary* resources that should continue to be employed to meet testing needs.

Which Facilities Should NASA Keep in Its Minimum Set?

Four factors should be used to assess which NASA facilities constitute the minimum set of strategically important facilities: alignment with national needs, technical competitiveness, redundancy, and usage. If a facility does not align with national needs, it should not be in the minimum set. However, simple alignment should not be the only deciding factor for inclusion, since such a factor does not consider other factors such as technical competitiveness and excess capacity.

The facilities that do not belong in the minimum set are those that, despite their alignment with national needs, are weakly competitive, redundant, and poorly utilized. The loss of these facilities will not strategically degrade the country's testing ability. Since remaining facilities that align with national needs cost anywhere from hundreds of millions to billions of dollars (in contrast with their operating budgets, which are in the millions) and take about 10 years to construct, they should be maintained in the minimum set.

Finally, NASA should not at this point leave out all facilities technically similar to those in the DoD. As we discussed above, further analysis of technical, cost, and availability issues are required to determine whether facility consolidation and right-sizing across NASA and AEDC would provide a net savings to the government.

What Should NASA Do with Its Low-Utilization Facilities?

Utilization data are unreliable beyond a few months into the future because programs do not have stable, quantitative testing plans more than a few months to a few years out. Also, revisions are common even within these plans. Utilization is cyclic, since new aeronautic programs start and end in multiyear cycles, and many mid-term strategic plans tend to change after a few years because of transforming priorities and technical breakthroughs. As noted above, since facilities cost anywhere from hundreds of millions to billions of dollars (in contrast to their operating budgets, which are in the millions) and take about 10 years to construct, we should not simply close facilities based on unreliable utilization data.

Mothballing an important facility is preferred to abandonment, but *mothballing involves the loss of workforce expertise required to safely and effectively operate such complex facilities.* Thus, mothballing is not an effective solution to dealing with long periods of low utilization, putting facilities at risk.

Utilization data are only one (nonexclusive) factor in determining what facilities to keep in the minimum set. In particular, utilization helps to decide what to do with redundant facilities. Thus, poorly utilized facilities should be reassessed for strategic, long-term needs rather than eliminated out of hand; only those that do not survive that assessment are candidates for closure or mothballing.

Finally, keeping all facilities open regardless of utilization is too extreme. Utilization is a useful channel of information that focuses attention on assessing strategic need and identifies management challenges to keep important facilities healthy for future needs. The result of a strategic reassessment of low-utilization facilities may indeed determine that a facility is no longer needed.

How Should NASA's Facilities Be Funded?

Setting the testing prices to cover all costs is not recommended because it can discourage use and endanger strategic facilities. This approach does give users more information about the full costs for conducting their tests at a facility. If this cost is too high, users can respond by seeking an alternative source of services if it is available; alternatively, users may avoid important testing or test in inferior facilities and obtain degraded or even misleading data. The approach would be a good option outcome if the alternative facilities are a better value over the long term and strategically important resources are retained. Unfortunately, this approach leads to poor outcomes if NASA is a better long-term value but low near-term utilizations and resulting higher near-term prices mask the long-term value. The approach is also bad when the remaining users cannot afford the costs to keep open strategic facilities needed in the long term.

Alternatively, setting the prices for user tests to zero closes one channel of information to users about the costs they are imposing on the infrastructure, can encourage overuse, and, therefore, can cause limits on the availability of funding. Thus, paying all costs through institutional funding so testing is free to users is not recommended.

NASA leadership should, therefore, **identify financial support concepts to keep its minimum set of facilities healthy for the good of the country.** It appears reasonable to ask users to pay for the costs *associated with their tests* (i.e., to pay for the short-term benefits), but forcing them to pay *all* operating costs (including long-term priorities such as the costs for facility time they are not using) through FCR *direct test pricing* (as is done at Ames) will further discourage use and endanger strategic facilities by causing wide, unpredictable price fluctuations in a world where government and commercial testing budgets are under pressure and are set years in advance.

Shared support options include (1) current internal NASA shared support mechanisms (e.g., those employed by Langley and Glenn), (2) a new budget item, and (3) a new testing technology research program. While the determination of the best shared support option is beyond the scope of this study, it is clear that shared funding support needs to be identified to preserve important facilities at Ames. NASA headquarters should ensure that FCR policies do not endanger the first option while determining the best long-term solution. NASA should also explore ways of sharing support with external entities through collaboration mechanisms such as consortia or retainers.

It is unclear how NASA will implement FCR of its test facility operating costs and whether the shared support taxing mechanisms currently in place at Langley and Glenn will be disallowed under the full-cost initiative.

Keep Perspective on the Costs Involved

It is important to retain perspective on the magnitude of NASA's WT/PT facility costs relative to the investment value of the aerospace vehicles they enable or support. The size of the overall NASA test facility operating budget is relatively small compared to the benefits it enables by ensuring that strategically important tools remain available to aerospace RDT&E developments important to this country for national security and the commercial industrial base. **While the approximately $125–130 million WT/PT facilities operating cost is a significant sum, it pales in significance to NASA's overall budget of about $15,000 million[3] and the $32,000–58,000 million the United States invests in aerospace RDT&E each year.** NASA should continue to closely reassess WT/PT needs and ensure that excesses are not present. However, the agency should keep in mind the connection between these costs and the benefits they accrue. Engineering practices indicate that both the short- and long-term benefits are worth the cost in terms of the vehicles they enable, the optimization gains, and the reductions in risk to performance guarantees—even if short-term budgets are not currently sized to support the long-term benefits.

What Future Investments Should Be Made?

While more than three-fourths of NASA's WT/PT facilities were judged competitive and effective with state-of-the-art requirements, there is room for improvement. The study identified a number of investment options that are warranted.

The significant, $128 million BMAR at NASA facilities should be eliminated to keep the facilities in the minimum set technically competitive with state-of-the-art requirements.

Various upgrades to existing facilities are needed to address identified challenges and to regain lost capabilities. The most readily identifiable major investment need is associated with the hypersonic vehicle programs. Serious research challenges in hypersonic air-breathing propulsion may require new facilities and test approaches for break-

---

[3] NASA 2002 Initial Operating Plan budget.

throughs to occur. This will require research in test techniques to understand how to address these testing needs and ultimately to look at the viability of hypersonic propulsion concepts being explored.

Another investment area is in high-productivity, high-Rn subsonic and transonic facilities. There certainly are areas where the nation is falling behind state-of-the-art capabilities and requirements. The Ames 12-Foot served the nation well in past decades but is now inferior to the QinetiQ 5-Metre in the United Kingdom in some aspects. Also, the NTF is a generation behind the European Transonic Windtunnel. However, the users surveyed had difficulty supporting the idea of a renewed pursuit of the facilities proposed in the National Facility Study[4]—not on technical research or programmatic risk-reduction needs but because the tight fiscal constraints that prevented the acquisition of these expensive facilities ($2–3 billion) in the late 1990s remain today in their programs. Thus, this investment option is technically warranted, but it is unclear how to pursue it.

Finally, the study has identified significant progress and utility from CFD; thus, we recommend continued investment in this research area.

Additional Recommendations

Beyond the options and recommendations discussed above in the table, we offer the following recommendations.

Focus on Specific Tunnels Requiring Attention
Financial shared support is most critical right now for the facilities at Ames: the 12-Foot, NFAC, and the 11-Foot WTs.

Until an alternative domestic high-Rn subsonic capability can be identified, the Ames 12-Foot Pressure Wind Tunnel should be retained in a quality mothball condition. The tunnel should *not* be abandoned at the end of FY2004 as planned unless an *alternative*

---

[4] *National Facilities Study* (1994).

domestic facility is identified in advance or an *explicit policy decision is made that it is in the interest of the nation to rely on foreign test facilities* (e.g., the QinetiQ 5-Metre tunnel in the United Kingdom) for all high-Rn subsonic testing.

The NFAC is strategically important, especially for the rotorcraft industry, and needs immediate financial support to prevent the facility from abandonment at the end of FY2004.

The Ames 11-Foot High-Rn Transonic facility currently provides excess capacity, but NASA should work with the DoD to establish long-term access to and clarified pricing for the AEDC 16T before any consideration is made to remove the Ames 11-Foot from the minimum set of needed facilities.

Other facilities with unhealthy ratings include the Langley Spin Tunnel, the Glenn 8×6-Foot Transonic Propulsion Tunnel, and the Ames 9×7-Foot Supersonic Tunnel. NASA leadership should further examine the health issues surrounding these facilities and develop methods to keep them viable (unless, in the case of the Langley Spin Tunnel, a decision is made to transfer responsibility to the Air Force).

Continue to Explore Options to Preserve the Workforce

Finally, while our principal study focus has been on the test facilities themselves, they are useless without trained personnel to operate them. Managing personnel issues in an organization in decline—the situation NASA faces in aeronautical testing—is also challenging. The approach NASA (and DoD at AEDC) has taken to address these personnel issues is based on crew sharing and transitioning the workforce from civil service to contractors. Crew sharing across multiple facilities has been an effective way to balance utilization variances across a wider facility base when cross-training is implemented, skills are preserved and honed in actual tests a few times a year, and standard control equipment are installed. Crew outsourcing is another consideration requiring further analysis but has one critical caveat: The civil-service cadre that provides the leadership and corporate memory for the test facility enterprise must be sustained at a critical mass to ensure that the government can provide sound direction and investment decisions and is a wise buyer of contractor services. Stabilizing

NASA's institutional support for test facilities will help ensure that today's dedicated and competent workforce will be able to pass its skills on to future generations.

Alternative Management Options

Alternative management approaches such as the UK MOD's privatization of T&E facility operations could be considered by NASA in the future for operating its facilities, but shared support would still be required, and costs and risks must be better understood. In any event, NASA should **continue to pursue full-cost accounting**[5] to inform current and future facility management decisions.

Bottom Line

NASA has three basic interrelated WT/PT facilities management issues to address: risk management, resource management, and intra- and interorganizational collaboration. NASA could continue to operate these facilities in a decentralized manner with increasing emphasis on FCR pricing, but this path risks uncoordinated consolidations and closures, endangering the nation's aeronautic RDT&E capabilities and the national security and industrial bases that rely on these capabilities. NASA could budget to shore up these facilities on its own, but dedicating resources to sustain these facilities would likely be very difficult given current budgetary pressures. NASA could collaborate more extensively with the DoD, industry, and possibly even foreign entities to share capabilities and expenses in a declining global RDT&E environment; however, the agency would need to address different priorities and organizational barriers with added vigor and support in a trusted and reliable fashion. The risks are real. Resources will need to be committed. Collaboration is possible and has been explored somewhat but not formalized.

---

[5] As opposed to FCR through pricing.

Congress has expressed interest in collaboration between NASA and the DoD.[6] NASA and the DoD (through their alliance NATA) have made some progress in collaboration,[7] but NASA's recent unilateral decision to close two facilities at Ames without high-level DoD review shows that progress has been spotty. Some in industry have expressed an interest in exploring collaborative arrangements with NASA and hope that this study will reveal to others in industry the risks to NASA's facilities and the need for industry to coordinate its consolidations with those of NASA and the DoD. Our study provides insights into the problem but offers only glimpses into the wider possibilities and issues surrounding broader collaboration.

In conclusion, NASA has played critical roles in advancing the aeronautic capabilities of the United States and continues to have unique skills and facilities important to the nation across the military, commercial, and space sectors—in terms of both research and support of our ability to learn about and benefit from advanced aeronautic concepts. Major wind tunnel and propulsion test facilities continue to have a prominent position in supporting these objectives. Unless NASA, in collaboration with the DoD, addresses specific deficiencies, investment needs, budgetary difficulties, and collaborative possibilities, the United States will face a real risk of losing the competitive aeronautics advantage it has enjoyed for decades.

---

[6] See, for example, the GAO report on NASA and DoD cooperation (*Aerospace Testing: Promise of Closer NASA/DoD Cooperation Remains Largely Unfulfilled,* 1998).

[7] For example, NATA has produced a number of joint NASA and DoD consolidation studies. See NATA (2001a; 2001b; 2002).

Bibliography

AAI Corporation, "Shadow 200 TUAV Program Finishing Successful Initial Tests; Program Moving to Next Phase Flight Testing at Ft. Huachuca," Hunt Valley, Md., October 16, 2000. Online at www. shadowtuav.com/newsreleases.html (last accessed April 2004).

Airborne Laser, "Modified Airborne Laser Aircraft Takes Successful First Flight from Boeing Flight Line in Wichita," St. Louis, Mo., July 18, 2002. Online at www.airbornelaser.com/news/2002/071802.html (last accessed April 2004).

Aeschliman, D. P., and W. L. Oberkampf, "Experimental Methodology for Computational Fluid Dynamics Code Validation," *AIAA Journal*, Vol. 36, No. 5, May 1998, pp. 733–741.

AFOSR—*see* Air Force Office of Scientific Research.

Air Force Office of Scientific Research, *Research Highlights*, July/August 2002.

Antón, Philip S., Dana J. Johnson, Michael Block, Michael Brown, Jeffrey Drezner, James Dryden, Eugene C. Gritton, Tom Hamilton, Thor Hogan, Richard Mesic, Deborah Peetz, Raj Raman, Paul Steinberg, Joe Strong, and William Trimble, *Wind Tunnel and Propulsion Test Facilities: Supporting Analyses to an Assessment of NASA's Capabilities to Serve National Needs,* Santa Monica, Calif.: RAND Corporation, TR-134-NASA/OSD, 2004 [companion document to this monograph; referred to as Antón, 2004(TR)].

Baals, Donald D., and William R. Corliss, *Wind Tunnels of NASA*, Washington, D.C.: NASA History Office, SP-440, 1981.

Batchelor, G. K., *An Introduction to Fluid Dynamics*, Cambridge, United Kingdom: Cambridge University Press, 1967.

Beach, H. L., Jr., and J. V. Bolino, *National Planning for Aeronautical Test Facilities*, Reston, Va.: American Institute of Aeronautics and Astronautics, 94-2474, 1994.

Bergquist, George W., Victor W. Hammond, D. V. Schnurr, S. Harry Meiselman, David S. Jewett, Henry D. McGlade, Clayford T. Everett, John D. Alexander, and John W. Cooley, *A Study of Funding Policy for Major Test & Evaluation Support Activities*, Office of the Assistant Secretary of Defense, Washington, D.C., April 25, 1972.

Boeing Corporation, "X-36 Tailless Research Aircraft Completes First Flight," St. Louis, Mo., May 21, 1997. Online at www.boeing.com/news/releases/mdc/97-118.html (last accessed April 2004).

_____, "Boeing Completes JSF X-32A Flight Testing," St. Louis, Mo., February 5, 2001. Online at www.boeing.com/news/releases/2001/q1/news_release_010205n.htm (last accessed April 2004).

_____, "Boeing X-45A Unmanned Combat Air Vehicle Begins Flight Testing," St. Louis, Mo., May 23, 2002a. Online at www.boeing.com/news/releases/2002/q2/nr_020523m.html (last accessed April 2004).

_____, "Boeing Unveils Bird of Prey Stealth Technology Demonstrator," St. Louis, Mo., October 18, 2002b. Online at www.boeing.com/news/releases/2002/q4/nr_021018m.html (last accessed April 2004).

_____, "Boeing 7E7 Dreamliner Will Provide New Solutions for Airlines, Passengers," 2003. Online at www.boeing.com/commercial/7e7/background.html (last accessed April 2004).

Bonneville Power Administration, "Transmission Business Line Unofficial Glossary of Terms," www2.transmission.bpa.gov/Business/Business_with_TBL/GLOSSARY2003.pdf (last accessed April 2004).

Chambers, Joseph R., *Concept to Reality: Contributions of the Langley Research Center to U.S. Civil Aircraft of the 1990s*, Hampton, Va.: NASA Langley Research Center, NASA/SP-2003-4529, June 2003. Online at http://techreports.larc.nasa.gov/ltrs/dublincore/2003/spec/NASA-2003-sp4529.html (last accessed April 2004).

_____, *Partners in Freedom: Contributions of the Langley Research Center to U.S. Military Aircraft of the 1990's*, Monographs in Aerospace History

Number 19, The NASA History Series, NASA History Division, NASA/SP-2000-4519, October 2000. Online at http://techreports.larc.nasa.gov/ltrs/dublincore/2000/spec/NASA-2000-sp4519.html (accessed April 2004).

Corliss, Bryan, "767 Tanker Takes to the Sky," *Everett [Wash.] Herald*, June 28, 2003.

Coutts, K. J., "Average Cost Pricing," in J. Eatwell, M. Milgate, and Peter Newman, eds., *The New Palgrave: a Dictionary of Economics*, Vol. 1, New York: Stockton Press, 1998, pp. 158–159.

Crook, A., "Skin-Friction Estimation at High Reynolds Numbers and Reynolds-Number Effects for Transport Aircraft," in Center for Turbulence Research, *Annual Research Briefs 2002*, Stanford, Calif., 2002, pp. 427–438.

Cruz, J. R., M. Kandis, and A. Witkowski, "Opening Loads Analyses for Various Disk-Gap-Band Parachutes," AIAA 2003-2131, 17th AIAA Aerodynamic Decelerator Systems Technology Conference and Seminar, Monterey, Calif., May 19–22, 2003. Online at http://techreports.larc.nasa.gov/ltrs/PDF/2003/aiaa/NASA-aiaa-2003-2131.pdf (last accessed April 2004).

Curtin, M. M., D. R. Bogue, D. Om, S. M. B. Rivers, O. C. Pendergraft, Jr., and R. A. Wahls, "Investigation of Transonic Reynolds Number Scaling Methods on a Twin-Engine Transport," AIAA 2002-0420, 40th AIAA Aerospace Sciences Meeting and Exhibit, Reno, Nev., January 14–17, 2002.

Customer Education Working Group, Audience/Messages Subcommittee, *Recommendations for Customer Education Plan About Retail Electric Competition*, submitted to the Arizona Corporation Commission, July 31, 1998.

Drezner, Jeffrey A., Giles K. Smith, Lucille E. Horgan, J. Curt Rogers, and Rachel Schmidt, *Maintaining Future Military Aircraft Design Capability*, Santa Monica, Calif.: RAND Corporation, R-4199-AF, 1992.

Drezner, Jeffrey A., and Robert S. Leonard, *Innovative Development: Global Hawk and DarkStar—Flight Test in the HAE UAV ACTD Program*, Santa Monica, Calif.: RAND Corporation, MR-1475-AF, 2002.

FAA—*see* Federal Aviation Administration.

Federal Aviation Administration, Associate Administrator for Commercial Space Transportation, *2002 US Commercial Space Transportation Developments and Concepts: Vehicles, Technologies, and Spaceports*, January 2002.

General Atomics, "Predator UAV Marks 50,000 Flight Hours," San Diego, Calif., October 30, 2002. Online at www.ga.com/news/50000_flight.html (last accessed April 2004).

Giunta, Anthony A., Steven F. Wojtkiewicz, Jr., and Michael S. Eldred, "Overview of Modern Design of Experiments Methods for Computational Simulations," AIAA 2003-0649, 41st Aerospace Sciences Meeting and Exhibit, Reno, Nev., January 6–9, 2003.

GlobalSecurity.org, "AV-8 Harrier Variants," July 31, 2002a (last modified), www.globalsecurity.org/military/systems/aircraft/av-8-variants.htm (last accessed April 2004).

_____, "Outrider Tactical UAV," November 19, 2002b (last modified), www.globalsecurity.org/intell/systems/outrider.htm (last accessed April 2004).

Lockheed Martin Aeronautics Company, "Lockheed Martin X-35 Flight Test Update No. 8," Marietta, Ga., February 5, 2001. Online at www.lmtas.com/news/programnews/combat_air/x35/x35_01/x35pr010205.html (last accessed April 2004).

_____, "Manufacturing Begins on Lockheed Martin F-35 Airframe," Marietta, Ga., November 10, 2003. Online at www.lockheedmartin.com/wms/findPage.do?dsp=fec&ci=13498&rsbci=0&fti=112&ti=0&sc=400 (last accessed April 2004).

Lorell, Mark A., and Hugh P. Levaux, *The Cutting Edge: A Half Century of U.S. Fighter Aircraft R&D*, Santa Monica, Calif.: RAND Corporation, MR-939-AF, 1998.

Lyles, General Lester L., Commander, Air Force Materiel Command, "Statement to the Commission on the Future of the United States Aerospace Industry, Research, Development, Test and Evaluation Infrastructure Panel," August 22, 2002. Online at www.aerospacecommission.gov/082202testimony/lyles.doc (last accessed April 2004).

Mack, M. D., and J. H. McMasters, "High Reynolds Number Testing in Support of Transport Airplane Development," 92-3982, 17th AIAA

Aerospace Ground Testing Conference, Nashville, Tenn., July 6–8, 1992.

Mars: Dead or Alive, dir. Marc Davis, PBS documentary, 2004.

NASA—*see* National Aeronautics and Space Administration.

NATA—*see* National Aeronautics Test Alliance.

The National Aeronautics and Space Act of 1958, 72 Stat. 426, 42 U.S.C. 2451 et seq., Pub. L. No. 85-568, as amended. Online at www.hq.nasa. gov/ogc/spaceact.html (last accessed April 2004).

National Aeronautics and Space Administration, NASA Policy Directive 9080.1E, "Review, Approval, and Imposition of User Charges," effective date April 27, 1999 (expiration date April 27, 2004).

_____, *Full Cost Initiative Agencywide Implementation Guide*, February 1999.

_____, *Aerospace Technology Enterprise Strategic Plan*, April 2001. Online at www.aerospace.nasa.gov/library/enterprise.htm (last accessed May 2004).

_____, *NASA FY2003 Budget Amendment: Integrated Space Transportation Plan International Space Station*, November 12, 2002.

_____, *NASA 2003 Strategic Plan*, NP-2003-01-298-HQ, 2003.

_____, *FY 2004 Initial Operating Plan*, March 10, 2004. Online at www. nasa.gov/pdf/56694main_Op_Plan_12032004.pdf (last accessed May 2004).

National Aeronautics and Space Administration, Office of Aerospace Technology, *The NASA Aeronautics Blueprint—Toward a Bold New Era of Aviation*, NP-2002-04-283-HQ, February 2002. Online at www. aerospace.nasa.gov/aero_blueprint/index.html (last accessed November 26, 2003).

_____, *Turning Goals into Reality: Aerospace Technology Enterprise Annual Report 2001*, 2001.

National Aeronautics and Space Administration, Office of the Chief Financial Officer, *NASA FY 2003 Congressional Budget Request—Aerospace Technology*, n.d. Online at http://ifmp.nasa.gov/codeb/budget2003/15-Aerospace_Technology.pdf (last accessed April 2004).

National Aeronautics Test Alliance, *National Aeronautics Testing Alliance Transonic Wind Tunnel Study*, August 29–30, 2001a.

_____, *Supersonic Wind Tunnel Study*, October 1, 2001b.

_____, *Air Breathing Engines Test Facility Study*, March 12, 2002.

National Facilities Study: Facility Study Office on the National Wind Tunnel Complex Final Report, April 29, 1994.

National Research Council, Committee on Breakthrough Technology for Commercial Supersonic Aircraft, *Commercial Supersonic Technology: The Way Ahead*, Washington, D.C.: National Academy Press, 2001.

Northrop Grumman, "Northrop Grumman's X-47A Pegasus First Flight Achieves Milestone in Autonomous Control," Los Angeles, February 23, 2003. Online at www.irconnect.com/noc/pages/news_releases.mhtml?d= 37039 (accessed April 2004).

Oberkampf, W. L., and D. P. Aeschliman, "Joint Computational/ Experimental Aerodynamics Research on a Hypersonic Vehicle—Part 1: Experimental Results," *AIAA Journal*, Vol. 30, No. 8, August 1992, pp. 2000–2009.

Oberkampf, W. L., and F. G. Blottner, "Issues in Computational Fluid Dynamics Code Verification and Validation," *AIAA Journal*, Vol. 36 No. 5, May 1998, pp. 687–695.

Ortiz, Catalina, "NASA Centers Test Parachutes for Mars Rovers," *FO Outlook*, May 2003. Online at http://windtunnels.arc.nasa.gov/pics/ Outlook/May_03_Outlook.pdf (last accessed April 2004).

Pioneer UAV Inc., The Pioneer [UAV] System: Operational History, www. puav.com/pioneer_history.asp (last accessed April 2004).

Rahaim, Christopher P., William L. Oberkampf, Raymond R. Cosner, and Daniel F. Domink, "AIAA Committee on Standards for Computational Fluid Dynamics: Status and Plans," AIAA 2003-0844, 41st Aerospace Sciences Meeting and Exhibit, Reno, Nev., January 6–9, 2003.

Raytheon Corporation, "First Flight of Production T-6A Texan II Trainer," in *Raytheon 1998 Annual Report*, Waltham, Mass., 1998. Online at www.raytheon.com/finance/1998/annrpt/ac04.htm (last accessed April 2004).

Rutherford, Donald, *Dictionary of Economics*, New York: Routledge, 1992.

Smith, Bruce A., "Boeing to Set Sonic Cruiser Design in 2004," *Aviation Week & Space Technology*, February 11, 2004. Online at www.aviation now.com/content/ncof/ncf_n63.htm (last accessed May 2004).

Streett, C. L., "Designing a Hybrid Laminar-Flow Control Experiment: The CFD Experiment Connection," AIAA 2003-0977, 41st Aerospace Sciences Meeting and Exhibit, Reno, Nev., January 6–9, 2003.

Streett, C. L., D. P. Lockard, B. A. Singer, M. R. Horrami, and M. M. Choudhari, "In Search of the Physics: The Interplay of Experiment and Computation in Airframe Noise Research; Flap-Edge Noise," AIAA 2003-0979, 41st Aerospace Sciences Meeting and Exhibit, Reno, Nev., January 6–9, 2003.

U.S. Department of Defense, *DoD Aeronautical Test Facilities Assessment: DoD's Future Aeronautical Development Program Needs for Wind Tunnel Testing and Computational Fluid Dynamics*, March 1997.

_____, *Joint Vision 2020*, June 2000. Online at www.dtic.mil/jointvision/jvpub2.htm (last accessed May 2004).

_____, DoD Directive 3200.11, "Major Range and Test Facility Base (MRTFB)," Office of the Secretary of Defense, Operational Test and Evaluation Directorate (DOT&E), May 1, 2002. Online at www.dtic.mil/whs/directives/corres/pdf/d320011_050102/d320011p.pdf (last accessed December 10, 2003).

_____, *Unmanned Aerial Vehicles Roadmap 2002–2027*, Office of the Secretary of Defense, December 2002.

_____, "Reimbursable Operations, Policy and Procedures," Volume 11A, in *DoD Financial Management Regulation*, DoD 7000.14-R, Under Secretary of Defense (Comptroller), updated April 8, 2003. Online at www.defenselink.mil/comptroller/fmr/11a/index.html (last accessed April 2004).

U.S. General Accounting Office, *Aerospace Testing: Promise of Closer NASA/DOD Cooperation Remains Largely Unfulfilled*, GAO/NSIAD-98-52, March 1998.

U.S. Senate, *Departments of Veterans Affairs and Housing and Urban Development, and Independent Agencies Appropriations Bill*, Senate Report 107-43 from the Committee on Appropriations, 107th Congress, 1st Session, July 20, 2001.

_____, *Aeronautics Research and Development Revitalization Act of 2003*, Bill S.309, 108th Congress, 1st Session, introduced in the U.S. Senate on February 5, 2003.

Wahls, R. A., *The National Transonic Facility: A Research Retrospective*, Reston, Va.: American Institute of Aeronautics and Astronautics, AIAA-2001-0754, January 2001.

Walker, M. M., and W. L. Oberkampf, "Joint Computational/ Experimental Aerodynamics Research on a Hypersonic Vehicle—Part 2: Computational Results," *AIAA Journal*, Vol. 30, No. 8, August 1992, pp. 2000–2009.

Walker, Robert S., F. Whitten Peters, Buzz Aldrin, Edward M. Bolen, R. Thomas Buffenbarger, John W. Douglass, Tillie K. Fowler, John J. Hamre, William Schneider, Jr., Robert J. Stevens, Neil deGrasse Tyson, and Heidi R. Wood, *Commission on the Future of the United States Aerospace Industry: Final Report*, November 2002.

White, B. W., "From Chutes to Wings: Coming Home from Space," Suite101.com, December 17, 2000, www.suite101.com/article.cfm/ residence_space/55346 (last accessed April 2004).